IMAGES
of America

PLYMOUTH AND NORTHWESTERN AMADOR COUNTY

IMAGES
of America

PLYMOUTH AND NORTHWESTERN AMADOR COUNTY

Deborah Coleen Cook and Amy Elizabeth Champ

ARCADIA
PUBLISHING

Published by Arcadia Publishing
Charleston, South Carolina

Printed in the United States of America

Library of Congress Control Number: 2022938575

For all general information, please contact Arcadia Publishing:
Telephone 843-853-2070
Fax 843-853-0044
E-mail sales@arcadiapublishing.com
For customer service and orders:
Toll-Free 1-888-313-2665

Visit us on the Internet at www.arcadiapublishing.com

This book is dedicated to Johnny and Larry.
You left us much too soon and are greatly missed.

CONTENTS

ACKNOWLEDGMENTS

This book has been years in the making and would not have been possible without stories told to Deborah Cook by residents of Plymouth and the surrounding countryside. She would especially like to thank the Colburn family, who have been Plymouthians for over 80 years. Marie, the matriarch of the family, was always generous with her mid-century tales of the town during our evening visits on her front porch. Likewise, her children Marla, Nan, Gary, and Jon were always delighted to share their childhood memories of growing up in Plymouth. We would also like to thank our colleague and friend, author Elaine Zorbas. Her publications on Fiddletown and the Gold Rush Chinese community have been invaluable sources in compiling this volume. Our gratitude is also extended to Ella Emerson and her daughter Carol, Mary Cowan, and Linda Weber, for sharing their knowledge of the area with us. Our thanks also go out to Michael Spinetta for providing us with images from his private collection and taking time out of his busy schedule to take photographs for this volume. In addition, we would like to thank the County of Amador for the generous use of images from its historic photograph collection at the Amador County Archives.

INTRODUCTION

Beyond the rolling grasslands of the eastern Sacramento Valley, skirting the banks of the Cosumnes River, is a landscape that entices the visitor to return again and again. Take away the paved roads, streetlights, and other modern appurtenances and the nature of this region remains much as it was when visited in 1857 by a roving reporter for the *Volcano Weekly Ledger*. Riparian vegetation still grows thickly along the banks of the river. Annual grasses and wildflowers burst forth in bright hues on the hillsides. Then, as summer approaches, the flowers fade, and the grasses turn to a golden hue. The oaks that were once thick as forests in the lower regions are now represented by scatters of the leafy giants. Among them is the buckeye, which the native Miwok once harvested to supplement their acorn stores, still blooming along the roadsides. But gone are the miners standing knee deep in water and bent over rockers. Roadside stage stops have given way to hotels. Horse-drawn wagons and plows have been replaced with semitrucks and tractors. The never-ending sound of the stamp mills at the mines has been silenced for decades. And, although the dusty main streets of yesteryear have been paved over, the small-town ambiance and rural lifeways remain.

Settlement in the northwestern region of Amador County came on the heels of the thousands of miners who flocked to the region after John Marshall discovered gold in the tailrace of John Sutter's sawmill. Small mining camps such as Possum, Rich, and Cape Cod Bars sprung up along the Cosumnes River. Miners worked along Little Indian Creek at Puckerville. Within a decade, Yeomet and Bridgeport on the Cosumnes would be settled, and Drytown and Fiddletown were bustling hamlets with businesses established and families settled. Plymouth grew up around the trading post built there in 1857 by Joseph Chandler Williams. As the mining industry grew and more settlers moved into the surrounding hills, the town would thrive as an economic crossroads. The amalgamation of multiple claims into the Plymouth Consolidated Mine and the birth of a very profitable wine industry fueled the growth of northwestern Amador into the 20th century. Even with the setbacks of the Great Depression and the closing of the mines in 1942, the region continued to thrive.

Gone are the mining camps along the Cosumnes and the settlements of Yeomet and Enterprise. Drytown and Fiddletown are no longer the bustling towns they once were, and Bridgeport is now known as River Pines. New Chicago and Rancheria are but shadows of their past. Moving into the 21st century, the region has redefined itself. The historic towns and the world-renowned Shenandoah Valley wine industry has made this place a favorite for tourists. Infrastructure improvements and new housing developments as well as a Native American gaming facility soon to be built near Plymouth may change the face of this small town, but the values that were brought by the first families who settled here and their spirit of adventure remain.

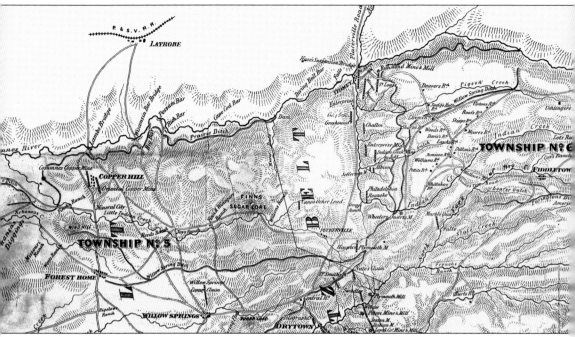

This section of the 1866 official map of Amador shows how well settled the northwestern region of the county was within two decades of the arrival of the first prospectors. Numerous towns were established from the Cosumnes River south to Drytown and from the Sacramento County line east to Fiddletown and Bridgeport. Many of the mining camps shown here no longer exist. (Authors' collection.)

One

EARLY EXPLORATION

INTO THE NATIVE LAND

In June 1542, Iberian maritime explorer Juan Rodriguez Cabrillo, under orders from conquistador Herman Cores, sailed north from Navidad, Mexico, past the Gulf of California, and dropped anchor in San Diego Bay. Nine days later he reached Santa Catalina Island, where he had his first encounter with California natives. He noted in his log, "A great crowd of armed Indians appeared." During his five-month voyage, Cabrillo recorded the name of a Chumash village along the coast and the nearby islands. This initial encounter between the Spaniards and Indians appears to have been peaceful. However, 200 years later in 1769, that relationship would alter drastically when Gaspar de Portola led a party north and established the Presidio of San Diego. He was accompanied by Fr. Junipero Serra, who founded the Basilica San Diego de Alcala, the first of 22 Catholic missions built along the California coast. Within a month, the Indians revolted and battled with the newcomers. An Indian boy died and three Spaniards were wounded. Unfortunately, this was but the first of many conflicts between the Indians and the Europeans. It was a portent of things to come.

Over the next 150 years, California Indians were enslaved, murdered, run off their land, and decimated by disease. When California acquired statehood in September 1850, the federal government appointed agents to oversee Indian affairs. Shortly thereafter, it was reported that there was an Indian uprising along the banks of the Cosumnes River, which eventually became the northern boundary of Amador County. Indian agent Oliver M. Wozencraft was dispatched to the area and instructed to sign treaties with the Indians there. His efforts, which included the involvement of some nefarious characters, was a fiasco for the Indians. However, despite this failure and the challenges the Indians have faced since then, the Northern Sierra Miwok, whose ancestral home included the region covered in this book, have continued to live here. The Ione Band of Miwok, who are descendants of the Northern Sierra Miwok and Nisenan peoples, have restored their first trust lands and will soon build a gaming facility near Plymouth.

Frontiersman Jedediah Strong Smith was the first white man to explore the upper reaches of the Cosumnes River in 1826 and interact with the Indians there, while searching for the legendary Buenaventura River that supposedly flowed from the Rocky Mountains to the Pacific Ocean. This is the only known contemporary image of him, which was drawn by a friend four years after Smith's death in 1831. (Courtesy Larry Cenotto.)

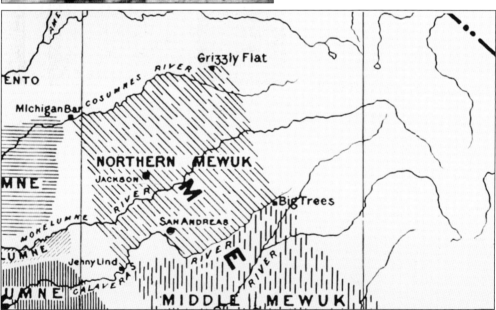

This 1910 map, compiled by famed naturalist and ethnographer C. Hart Merriam, shows the distribution of Native American Miwok people in the Sierra Nevada foothills. The Northern Miwok occupied an area in northwestern Amador County from Michigan Bar east to Grizzly Flat in El Dorado, with their northernmost boundary along the Cosumnes River. To the south, their territory extended to the Calaveras River. (Authors' collection.)

In 1848, Jared Dixon Sheldon, one of the first settlers on the Cosumnes River east of Sacramento, commissioned William Tecumseh Sherman (right) to conduct a survey of his vast land holdings. Sherman's survey took him and his party up the Cosumnes River into the area that is now within Amador County where he encountered several village sites of the Northern Miwok. It is probable that he visited the village of Yuleyumne, which was about a mile west of where Plymouth is situated (No. 16 on the map below). Other nearby Miwok villages included Omo (No. 17), at the future townsite of Omo Ranch, and No-ma (No. 18), where the mining camp of Indian Diggings would be established. This map, compiled in 1925 by anthropologist Alfred L. Kroeber, also shows the location of other Miwok villages to the south and west. (Right, courtesy National Portrait Gallery; below, courtesy University of California.)

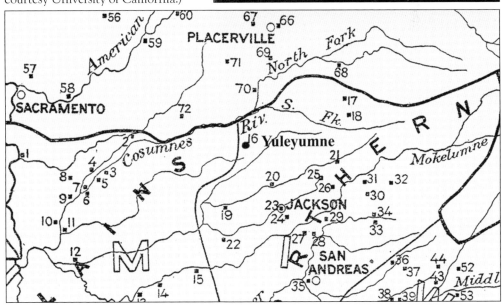

In 1850, Oliver M. Wozencraft (left) was appointed as a California Indian agent. The following year, he signed a treaty on behalf of the government with the Miwok along the Cosumnes River. However, that agreement was short lived. He was recalled from his position after he contracted with shady businessmen to distribute beef purchased by the federal government to the Indians. The best of the beef was sold off to other buyers and the dregs of the herd given to the Indians. Lawsuits were filed by various people involved in the matter. William R. Grimshaw (below) was called to testify in a suit filed by Samuel Norris against the government for payment of the beef he had purchased on their behalf to distribute to the Indians. (Left, courtesy San Bernardino Public Library; below, authors' collection.)

Franklin Walker arrived in California in 1850 and headed for the Cosumnes River, where he mined for a time and then set up a trading post. Walker testified at the hearing against Oliver Wozencraft, telling the inquisitors that the Indians thereabouts supported themselves satisfactorily by trading and they made a good deal of money mining alongside the white men. (Courtesy Larry Cenotto.)

Euro-American prejudice against Native Californians was promoted by the state's first governor, Peter Hardeman Burnett. His inaugural address was laced with language that supported the notion that the genocide of California Native Americans was inevitable. Burnett's prejudice was inculcated during his Tennessee childhood, his family having battled against the Cherokee. His gubernatorial platform toward Indians set the stage for state legislation that followed suit. (Courtesy Larry Cenotto.)

Fish was a vital part of the Miwok diet. A number of methods were used to catch or trap the fish. Dip nets, attached to a long pole, were used in deep holes in rivers. Seine nets, known as *yo'ho*; set nets (*yu'gu*); spears (*hotca*); and basket traps were also employed. Small fish in shallow waters were often caught by hand. (Courtesy Library of Congress.)

The ecosystem along the Cosumnes River, consisting of a mixture of oak woodland, riparian vegetation, and grassland habitats, provided ample food supplies for Native Americans. They hunted game, which included deer, elk, rabbits, and game birds. Nuts, seeds, and insects also provided protein to their diet. Other foods included berries, mushrooms, greens, roots, and bulbs. (Courtesy California Department of Fish and Wildlife.)

Miwok houses, called *u'macha*, were built over a shallow pit in the ground that ranged from 8 to 15 feet in diameter. Grapevines or small willow branches were woven around supporting cedar poles and covered with cedar bark. The photograph at right shows a replica *u'macha* at Indian Grinding Rock State Historic Park. The sweat lodge (below) was also a feature of the Miwok village. These low conical-shaped structures were used for spiritual purification ceremonies known as a "sweat," which is still practiced in present-day Miwok culture. The ceremony is led by elders who have been trained and know the traditional language, songs, and safety to be practiced during the ceremony. (Both, authors' collection.)

Miwok village sites also included a roundhouse used for ceremonial gatherings. They were constructed much like the houses but ranged from 40 to 50 feet across. Perimeter posts set into the ground, and heavy beams supported a conical roof covered in earth, thatch, and cedar bark. Inside was a fire pit maintained by a *wŭk'ppeŭ* (fire tender). In the 1904 image above, Chief Hunchup stands next to a roundhouse at a village site on the Cosumnes. Below is a view inside the ceremonial roundhouse at Indian Grinding Rock State Historic Park, which shows the intricate work that went into its construction. All of the materials used to build the massive structure were harvested from the local forest. The firepit can be seen in the center of the dugout floor. (Above, courtesy Bancroft Library; below, courtesy David Abercrombie.)

Dancers often performed during ceremonies held in the roundhouse. The traditional dance costume consisted of standard accoutrements with personal artistic touches added by each dancer. Typical features included flicker feather headbands, magpie feather crowns, feathered skirts, and a feather plume. These were accessorized with belts, necklaces, paint, earplugs, animal pelts, and nose sticks, among other things. (Courtesy Larry Cenotto.)

In the autumn of every year, the Miwok still celebrate the acorn harvest at festivals called Big Time. In Amador County, it is held at Indian Grinding Rock State Historic Park in Pine Grove. This statue of a Miwok man in traditional dress stands in the park near the visitor's center. (Courtesy Library of Congress.)

The Miwok consumed a variety of plant material to supplement their diet of wild game. Nuts and seeds were ground into meal utilizing a handheld stone pestle that was pressed into a stone cup or mortar. Cups were also formed in bedrock outcroppings. This mortar station, named Chaw'se, is near the town of Volcano at Indian Grinding Rock State Historic Park. (Authors' collection.)

Portable bowl mortars and pestles of various sizes could be transported between seasonal village sites. In addition to grinding nuts and seeds, they were often used to pulverize dried fish and meat, medicinal plants, and mineral paints. This 1904 photograph was taken at a village site on the Cosumnes River. (Courtesy Bancroft Library.)

After gathering acorns, the Miwok stored them in granaries constructed of small tree branches and twigs. The framework was then lined with grass and roofed with materials such as thatch or pine boughs. Although small insects could possibly access the crop, these crude cupboards on stilts kept animals such as squirrels and birds from pilfering the much-needed food staple. (Courtesy Bancroft Library.)

This 1947 photograph taken by the Bureau of Indian Affairs shows a different type of acorn granary. Although this method of storage is no longer employed as standard practice, the Miwok continue to harvest acorns for flour and other foods. Acorn soup has become popular in upscale restaurants across the country. (Courtesy US Department of the Interior.)

This photograph of an unidentified Miwok woman holding a basket was taken in 1924 by the famed Native American photographer Edward S. Curtis. Just like with modern flour sifters, she would have used this basket to sift acorn flour. The flour was an ingredient in bread, porridge, and soups. (Courtesy Library of Congress.)

The art of basket weaving has been practiced by native cultures around the world for thousands of years. In this photograph, an unidentified Miwok woman constructs a basket of what appears to be reeds or slivered lengths of bark. This traditional art is still practiced by Miwok women today. (Courtesy Larry Cenotto.)

Handwoven baskets were used by the Miwok for assorted purposes, from cooking and kitchen vessels to gathering foodstuffs, infant cradles, and for ceremonial purposes. These were constructed of various materials, including grass, twigs from digger pine, slough grass root, redbud bark, and fern and bracken roots. Artistic expressions were often incorporated into the design, including anthropomorphic figures, zoological figures, and geometric patterns, and often included bird feathers and beads. Baskets of different sizes and shapes were a staple in the Miwok kitchen. This photograph shows a variety of baskets. At top left is a small dipper basket. No. 2 is a parching basket used in cooking. No. 3 is a canoe-shaped basket. Baskets used for serving food and general receptacles are shown in figures 4, 5, and 7. No. 6 is a hemispherical basket. The tradition of basketmaking continues to be practiced by the Miwok, although most modern baskets are only for decorative purposes. (Authors' collection.)

Native American life changed drastically when Euro-Americans moved into California. In addition to many losing their traditional homelands, they struggled to hold on to their cultural identity and maintain traditions. Everything from clothing styles to housing and food changed over time. In the photograph at left, taken in 1872, Miwok women have adopted the practice of dressing in blouses and skirts; however, three of them continue to wear a traditional headband. In the 1920s image below, a Miwok woman has donned traditional ceremonial dress for an upcoming celebration. (Both, courtesy National Park Service.)

The Ione Band of Miwok, who received federal recognition for their tribe in 1994, are keeping their heritage alive by practicing traditional material culture and passing it along to younger generations. Basketry and beadwork are two of the traditional crafts that many women participate in. Ramona Dutschke, a well-known basket maker from Ione, is pictured at right with Huell Howser, a local television personality, at a Chaw'se celebration. Dutschke, who passed away in 2006, was a distinguished and much-loved elder of the Ione Band. For many years she taught basketmaking techniques and other aspects of the Miwok culture at Chaw'se. The Miwok also share their heritage with the community. Below, a booth at the Amador County Fair displays a variety of traditional clothing and artifacts. (Both, courtesy Amador County Archives.)

The illustration above by Amador County artist Rand Huggett, titled *1848, Indians at Creek with Miners*, is a perfect depiction of the culture clash that occurred at the onset of the California Gold Rush. Members of the Ione Band of Miwok, numbering over 700, are descendants of the Native Americans who occupied areas in northwestern Amador County. In 2020, they announced that the Bureau of Indian Affairs placed land into federal trust for the tribe, ending a two-decades long legal battle to have land restored to them. The property is on California State Route 49 at the western edge of Plymouth. The tribe has established an office in Plymouth and plans to build a gaming facility on the trust lands. (Above, courtesy Rand Huggett; below, courtesy Mike Spinetta.)

Two

Rush to the Hills

Miners, Miners, Everywhere!

Few migrations in human history eclipse the one prompted by John Marshall's discovery of gold in the tailrace of John Sutter's sawmill on January 24, 1848. There, on the banks of the American River at Coloma, Marshall's find changed the course of world history. With the subsequent end to the Mexican-American War, California was now a possession of the United States. The stage was set, and the fledgling America was about to increase its riches by billions. As news of the lucky strike spread, adventurers from around the world rushed to the future Golden State, drawn by word of mouth and advertisements promising riches. Within six months, tents and shanties dotted the landscape along the Cosumnes River. Miners stood ankle deep in water bent over rockers and pans, while others dug ditches, built flumes, and tunneled into hillsides. In the hills and valleys away from the river, groups of miners worked in dry diggings. On October 19, 1848, the *Sacramento Transcript* ran a few lines about a find on the Cosumnes: "Gold in Quartz! A new and very rich discovery of gold in quartz has been found on the Cosumnes, about sixty miles from Stockton. The vein is said to be richer than any yet discovered in California!" The arrival of these fortune seekers changed both the natural and cultural landscapes. Barren land soon became occupied by clusters of cabins and tents, which eventually gave way to houses and businesses. Some of the early camps in northwestern Amador County were but brief stopping points marked on a map, while others flourished and eventually became towns like Plymouth, Drytown, and Fiddletown. Some miners became rich, others went broke, and some did neither. As William M'Collum, MD, who arrived in 1850 seeking his fortune, put it: "It was a wild, I may almost acknowledge, a hare-brained adventure, and yet it is over and past. . . . With California, and all that is in it, I quit even."

Within weeks of John Marshall's discovery, news of the find spread to the East Coast and Europe. Thousands of eager fortune seekers headed to the West Coast; many arrived by ship in San Francisco. Glidden & Williams owned a large fleet of ships that sailed mainly between Boston and San Francisco beginning in 1849. (Courtesy Larry Cenotto.)

The steamship *Senator*, also rigged as a two-masted schooner, was one of the first ships to set sail from the East to San Francisco in 1849. On her first voyage west, she picked up 200 adventurers at Panama who had crossed the Isthmus on foot. She also carried goods and passengers up the Sacramento River to Sutter's Fort, where they embarked for the mining regions. (Courtesy Mariner's Museum.)

San Francisco was but a sleepy coastal village at the beginning of 1848. A census conducted by school trustees tallied the entire population at around 850. Buildings numbered 200, including homes, boardinghouses, tenpin alleys, saloons, and two large hotels. Along the waterfront, warehouses had been erected and two wharves were being built. This photograph of a San Francisco street was taken in early 1849. (Courtesy Library of Congress.)

Within 12 months of the previous photograph, the population of San Francisco exploded. In 1849, numerous ships docked daily carrying hundreds of gold seekers. It is estimated that by June, 15,000 people had arrived. The influx continued into 1850, when this photograph was taken at Yerba Buena Cove. (Courtesy Library of Congress.)

As the Gold Rush progressed, fledgling California merchants jumped on the marketing bandwagon. Advertisements, such as this one for mining tools, romanticized the notion of becoming rich with very little effort. The bucolic atmosphere and sophisticated infrastructure pictured in this 1851 advertisement was far from the reality hopeful miners encountered when they reached their destination. (Courtesy Library of Congress.)

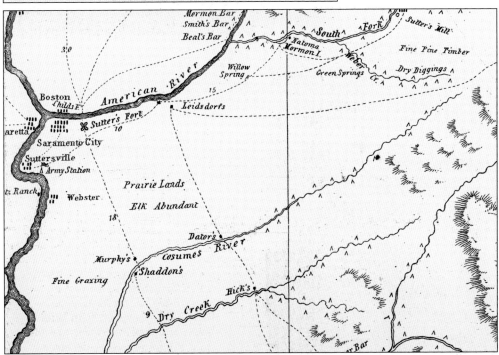

In 1850, William A. Jackson produced the "Map of the Gold District of California," a section of which is shown here. Near the center is the Cosumnes River, which forms the northern boundary of Amador County. Note the series of inverted Vs along the waterways, indicating mining activity. Mining along Dry Creek extended from Hick's trading post east to where Drytown is located. (Authors' collection.)

The Gold Rush was big news back east, and so the press sent reporters to cover it. In November 1849, Bayard Taylor, a correspondent for the *New York Tribune*, visited Drytown. He noted that it was "regularly laid out, 200 to 300 people, stores, butcher shops, monte tables, etc." Traveling west, he arrived at Willow Springs, "a log hut on the edge of a swamp." (Courtesy National Portrait Gallery.)

Lucius Fairchild, known for his prolific correspondence from the California goldfields, arrived in the mining camp of Big Bar on the Cosumnes River in November 1849. In letters home, he complained about the hard work of mining and the depraved living conditions the miners suffered: "Sleeping on the ground for over seven months, eating, sleeping, cooking, and setting on the ground in all weather." (Courtesy Library of Congress.)

Mining camps stretched along the Cosumnes River from Jared Sheldon's ranch at Sloughhouse east to Indian Diggings and beyond. At Michigan Bar, now in Sacramento County, a flurry of mining activity occurred. As county lines did not exist at the onset of the Gold Rush in 1848, many of the men who mined there would become citizens of northwestern Amador County. At its peak, Michigan Bar had over 2,000 people. Businesses included seven stores, five boardinghouses, two blacksmith shops, a butcher, a bakery, a drugstore, a tenpin alley, a livery stable, and a schoolhouse. A bird's-eye view of the town is shown above. Heath's store, pictured below, was built in the early 1850s. In the 1930s, it was purchased by Chinese miners, who tore it down to sift the earth beneath for gold. (Both, courtesy Library of Congress.)

Miners often worked in what were known as "dry diggings," locales where a natural water source was not available for pan washing or sluicing soil to remove the gold. In northwestern Amador County, a number of ditches were constructed to deliver water from the Cosumnes River to these dry mines. (Courtesy Library of Congress.)

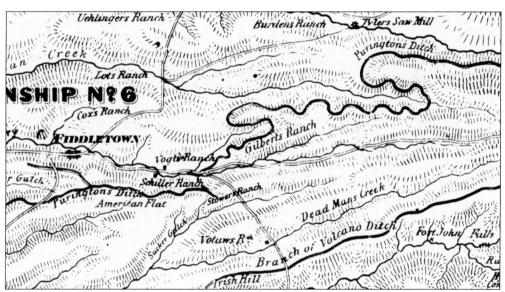

This section of the "1866 Official Map of Amador County" shows the Purinton Ditch, also known as the Cosumnes Ditch. It delivered water to the dry mines at American Flat and other nearby locales. The ditch was built by the Cosumnes River Mining and Ditching Company, owned by Columbus Purinton, his brother-in-law Benjamin Tyler, and William A. Davidson. (Courtesy Amador County Surveying.)

The Cosumnes River was the lifeblood of mining and settlement in northwestern Amador County. In addition to ditches, large wooden flumes were constructed to deliver water to mining claims and camps where no natural water source was available. In this rare 1850s photograph, ditch tenders are seen sitting on the Indian Creek flume of the Willow Springs Ditch, which drew water from the Cosumnes. (Courtesy Amador County Archives.)

One of the earliest mining settlements along the Cosumnes River was Indian Diggings. Although technically now in El Dorado County, many of the folks who owned property and did business here lived in the area that would become Amador County. The large flume seen in this early photograph of the town carried water from the river to the Purinton Ditch. (Courtesy Amador County Archives.)

The first mining to take place in the vicinity of Plymouth surrounded the settlement of Puckerville. Situated a mile west of the modern-day Plymouth, it straddled what is today called Old Sacramento Road. Here, the creeks and streams were dotted with mining claims, which are marked on this 1858 Government Land Office survey map as "old mines." (Courtesy California Division of Mines.)

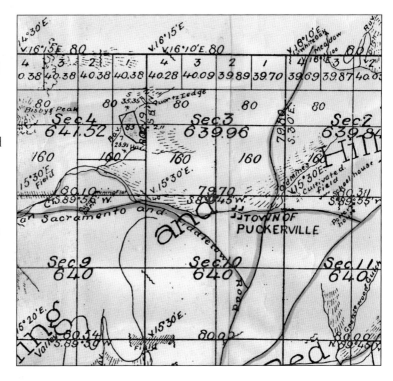

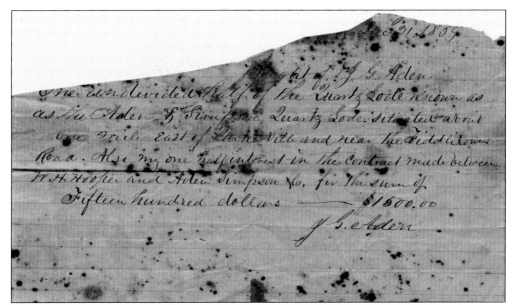

The earliest known mining claim to have been filed near the future townsite of Plymouth was taken up by Jeremiah Green Aden and Edward Simpson around 1849. It was south of the east end of Main Street. Ten years later, Aden sold his half of the Aden & Simpson Mine to John Hooper. Pictured here is the handwritten receipt of that transaction. (Courtesy Amador County Archives.)

The banks along Dry Creek were a hotbed of mining activity in 1848. Around this, the hamlet of Drytown, considered the oldest town in Amador County, grew quickly. When John Doble visited in 1852, he wrote in his diary that there were some 200 houses built, a post office, a school, and a Catholic church, and a stamp mill was running. (Courtesy Amador County Archives.)

The Seaton Mine was one of the largest producers at Drytown. It consisted of five linear-adjacent 120-foot claims. The mine was initially opened by George Seaton and Ben Ritchmeyer & Company. Their success later brought on wealthy investors, including newspaper magnate William Randolph Hearst and William Hooper, owner of the Plymouth Mine. (Courtesy Amador County Archives.)

Within a few short years of Marshall's gold discovery, Drytown's population grew to what is estimated to have been 10,000 residents. Mines, such as the ones pictured here, surrounded the town. They included the California, the Lost Hat, the Centennial, Chili Jim, the Crown Point, the Henry Clay, the Homestake, the Italian, the Maryland, and the Pocahontas, to name a few. (Courtesy Amador County Archives.)

Nearly all the early mining claims consisted of surface working of gravels along watercourses and dry diggings along ancient riverbeds. Miners soon found that there was also gold to be had in quartz rock and began to open what are known as hard rock mines. The Cosmopolitan Mine, pictured here, was one of the first of this type worked near Drytown. (Courtesy Amador County Archives.)

East of Drytown, near where New Chicago Road crosses Rancheria Creek, was the mining settlement of Lower Rancheria. History buffs will recognize this as the site of the notorious Rancheria Massacre that occurred on August 6, 1855, when six people were killed by a group of 13 Mexican nationals brandishing knives, guns, and an axe. That was the year these two photographs of the camp were taken. Mining began here in 1848. Within a year, commercial buildings and houses had been erected, and the banks of the creek were lined with miners, many of them Chilean and Mexican. By 1858, the camp was all but abandoned. (Both, authors' collection.)

American Flat, south of Fiddletown, was first mined in 1852 when claims were opened by numerous individuals and partnerships. Strikes were made in the rich gold-bearing gravels of an ancient riverbed. In 1853, the *Sacramento Daily Union* reported that Capt. Samuel Stowers and his son Thomas held the richest claim there. At 25 feet into their mine, they found "dirt actually glittering with gold." Nearby American Hill, French Flat, Conita Hill, Loafer Flat, Lone Hill, and Sucker Hill were also locations of rich strikes on the same gravel deposits. Pictured here are Canvin's Mine at American Flat (above) and miners working gravel at a dry digging like those worked in and around American Flat (below). (Both, courtesy Amador County Archives.)

When the news of the gold discovery reached East Asia, thousands of Chinese immigrated to California seeking *gum shan*—the "gold mountain." Prejudice against the newcomers prompted the legislature to pass a foreign miner's tax of $20 per month on non-citizens. This was repealed in 1851, but in 1852, another law was passed that charged them $3 per month. (Courtesy Amador County Archives.)

Fiddletown, like other communities in northwestern Amador County, began as a camp. Miners were drawn to the gravel beds in the vicinity, such as American Hill. This extremely rare 1850s photograph of the hamlet shows just how quickly towns rose around the mining claims as merchants and hoteliers moved in to accommodate the needs of the miners. (Courtesy Amador County Archives.)

Three

PLANTING ROOTS

SETTLERS FAR AND WIDE

It was the promise of riches that lured thousands to California during the Gold Rush; however, it was the fertile land, the climate, and a booming economy that kept many in the state and brought thousands more. Within two years of Marshall's discovery at Sutter's sawmill in Coloma, coastal ports like San Diego, Los Angeles, and San Francisco were fledgling cities, and Sacramento grew up around Sutter's Fort. Between 1850 and 1860, California's population grew from 92,597 to 379,994, an increase of 310.4 percent in just 10 years. By 1900, the population was counted at 1,485,053. Amador County was not formed until 1854; thus, accurate population numbers within the county boundaries are only available beginning in 1860. That year, 10,930 residents were counted. In 1900, it stood at 11,116. That number would see a huge increase during the first half of the 20th century.

Along the mother lode, some mining camps grew into towns, while others faded into obscurity. No longer were there just men living in tents and shanties. Houses, businesses, churches, and schools had been built. In northwestern Amador County, several towns were settled and growing. Indian Diggings, Bridgeport, and Yeomet were established along the Cosumnes River. South of the river was Puckerville, Plymouth, Drytown, and New Chicago—all of which grew up around the mines. To the west, roadhouses at Central House and Forest Home were built to accommodate travelers, and a small community was established at Willow Springs. Eastward into the hills, Fiddletown became a thriving community, and farms and ranches dotted the landscape of the Shenandoah Valley. Roads were established between the fledgling settlements, and express companies brought coaches to move passengers, goods, and money. To the north, in Eldorado County, the railroad came to Latrobe. Plymouth, being the closest town to the railhead, became a hub of business and commerce. In 1854, Amador became a county, and voting precincts and school districts were formalized. Where there was once barren land, farmers, merchants, seamstresses, hoteliers, doctors, tailors, storekeepers, miners, and others found a new life in the Golden State.

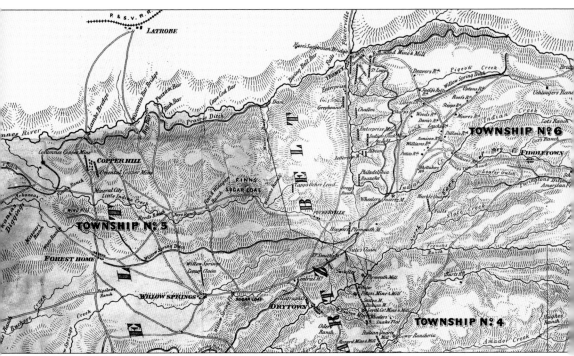

County boundaries did not exist when miners first arrived in California's Sierra Nevada foothills in 1849. It was not until 1850, when California was admitted to the Union, that 27 counties were formed. El Dorado and Calaveras were two of the original 27. The area that now comprises Amador County was at that time part of both El Dorado and Calaveras Counties, with Dry Creek the dividing line between the two. Then in 1854, Amador County was carved from them. By 1866, the first official map of Amador was completed; the northwestern region of the county was well settled, as can be seen in this detail of that map. Note that the locations of Fiddletown, Yeomet, Bridgeport, and Drytown are marked; however, Plymouth does not yet exist. Puckerville is also marked on the map to the west of where Plymouth would be established. (Courtesy Amador County Surveyor.)

The name Puckerville (or Pokerville) is often erroneously ascribed to Plymouth; however, Puckerville was situated on the Old Sacramento Road about a mile west of Plymouth. Miners first worked the ground along nearby Spring Creek in 1849. During its heyday in the early 1850s, Puckerville's population bloomed to over 200 people. This photograph of a dilapidated house at the village site was taken in 1934. (Courtesy Library of Congress.)

This c. 1880s image shows all that was left of Puckerville at that time. It is unknown what this building was used for originally, but it may have been owned by Fred Schairer and Wells Heuston. They operated a general store and stage stop at Puckerville, sold hay and grain, and rented rooms. The operation folded in 1863 upon the death of Heuston. (Courtesy Amador County Archives.)

Joseph Woolford, a native of England, was well known throughout northwestern Amador County. He arrived in California in 1862 and settled at Puckerville. Here, he farmed and plied the blacksmith trade and was contracted as such with both the Empire and Pacific mining companies. Woolford died a lifelong bachelor in 1911 and is buried in the Plymouth cemetery. (Courtesy Amador County Archives.)

Pictured here is the first schoolhouse built at Puckerville. The teacher could very well be W.O. Applebee, who was its first permanent teacher. The students pictured here could be the children of Eleazer Scoville Potter and his wife, Harriet, who were the first family to settle at the future townsite of Plymouth. (Courtesy Amador County Archives.)

This view of Plymouth was captured in the 1880s from the hill overlooking the town near the cemetery. Current residents may recognize the large two-story building at right as part of the Colburn home on Main Street. It was constructed in the 1870s as a meeting hall for the Caucasian Society, an anti-Chinese fraternity. (Authors' collection.)

In this photograph, another rare early view of Plymouth, one of a few hand-dug wells in the town can be seen. It was on a hill north of Main Street. Although the Arroyo Ditch brought water to town from the Cosumnes River, when the ditch dried up during the warm months, these wells were used by residents until indoor plumbing was installed. (Courtesy Amador County Archives.)

The Empire Store building is the oldest structure in Plymouth. It was built in 1857 by Joe Williams. When the Empire and Pacific mines were joined as Plymouth Consolidated, it operated as a company store. By 1891, it was taken over by Hiram Edgar Potter. In 1905, Bernard Levaggi purchased the building, and it is still owned by the Levaggi family. (Authors' collection.)

RESIDENCE AND LUMBER YARD OF **E.S. POTTER.**
PLYMOUTH, AMADOR C? CAL.

Eleazer Scovill Potter and his wife, Harriet, were the first permanent settlers in the area of what became Plymouth. The couple were married in 1858 and built their home at the future townsite. Eleazer and fellow homesteader Joseph Pinder sold off parcels of their property, which became the business and residential lots that formed Plymouth. (Courtesy Thompson & West.)

44

This schoolhouse, the second built at Plymouth, stood on the hill behind where Plymouth Park is now located. The land was donated to the Indian Creek School District by Eleazer and Harriet Potter. Children who previously attended classes at Puckerville now had a school close to home. This building replaced the first schoolhouse, which was burned to the ground by an arsonist. (Courtesy Amador County Archives.)

The Central Hotel in Plymouth once stood on the lot now occupied by city hall. The building was constructed by James Dingle in 1883. He and his wife, Emma, owned the business from that year until 1911. In 1888, James skipped town, and Emma continued to run the hotel until she sold it the following year to Robert and Anna Carroll. (Author's collection.)

This hotel at Central House was established in the early 1850s by Sanford P. Brown and Charles D. Smith. It stood where today's Highway 16 and Highway 49 intersect, on the lot now occupied by the Wellness Cottage. In 1863, John Grambart and Matthew Wells purchased the property; they later sold it to Elmer Rupley. The two girls pictured are sisters Audrey (left) and Melba Fancher. (Courtesy Amador County Archives.)

Farther west on the road to Sacramento stood the Forest Home Inn, approximately seven miles from Central House. This structure, no longer standing, was built in the early 1850s by Wellman and Frances Castle to serve as a stage stop. After the Willow Springs Ditch brought water to the dry diggings nearby in 1857, the proprietors provided room and board to miners. (Courtesy Amador County Archives.)

William Jennings, in partnership with a Mr. Richardson, built what was known as a "first class hotel" at Willow Springs, another traveler's stopping place on the road to Sloughhouse and Sacramento. Jennings later went on to work in several Amador County positions, including treasurer and supervisor of his district. It is not known what became of Richardson. (Courtesy Amador County Archives.)

Pictured here is the old Willow Springs schoolhouse that once stood on Greilich Road. Farmers and ranchers in the vicinity petitioned the Board of Supervisors for a school at Willow Springs in 1861. Parents footed the bill for the materials and built it. By 1863, the school had 61 pupils. The building burned in 1878 from a fire that spread from the wood stove. (Courtesy Amador County Archives.)

In this c. 1870s image, unidentified residents of Drytown took the opportunity to have their photograph taken in front of the town's butcher shop. The building pictured here was constructed in 1867, replacing the old butcher shop that once stood on the property. That building was destroyed in a fire that swept through the town in 1857. (Courtesy Amador County Archives.)

D.J. Bell's Exchange Hotel was the first establishment of its kind in Drytown. McWayne's general store, to the left of the hotel, was first owned by Calder Misner who went into partnership with Samuel B. Flanders and then Allen McWayne. When Misner moved to Alameda County, McWayne became the sole owner of the business. (Courtesy Amador County Archives.)

PACIFIC EXPRESS COMPANY.

LIST OF THE
PACIFIC
Express Company's
OFFICES.
San Francisco
Sacramento
Marysville
Nevada
Grass Valley
Mormon Island
Shasta
Placerville
Auburn
Coloma
Georgetown
Yankee Jim
Jackson
Ophir
Stockton
Sonora
Mokelumne Hill
Mariposa
Los Angeles
Portland, O
Crescent City, O
Port Orford, O
San Jose

No. 52

Drytown, April 14th 1856

Received of John Dillon ox package, addressed to Mrs. Ann Dillon, New York which we agree to forward to destination.
Value, One Hundred and Forty Dollars.
Freight and Insurance, $4 52/100 Paid

$140.

For the Pacific Express Co.

During the 1850s, the economy boomed in California, especially for those who were involved in mining. As the population grew, the Pacific Express Company began to operate on a regular basis in Amador County. Not only did it carry freight and provide passenger coach service, but it also handled cash money transfers, as shown on this receipt issued at Drytown in 1856. (Courtesy Amador County Archives.)

As the area grew, Plymouth became a hub of commerce and settlement in northwestern Amador County. Many folks did business or traveled to see friends and relatives in the Sacramento Valley and beyond. The Sacramento-Plymouth Stage provided passenger service between the two destinations, offering to drop off riders at several locations in the capital city. (Courtesy Amador County Archives.)

A short distance east of Drytown, on the Drytown–Amador City Road, once stood the village of New Chicago, which grew up around the Gover Mine. New Chicago was a boomtown for several decades. By the 1880s, it boasted all the amenities larger towns in the area had, but within 20 years, its population had dwindled and residents moved on. (Courtesy Amador County Archives.)

The Gover Mine, the lifeblood of New Chicago, had its beginnings in a questionable claim filed in the 1850s. Owners plugged along working it until 1877 when they hit a pay dirt bonanza. The old 10-stamp mill was replaced by a new larger one as the mine pumped out $35,000 to $40,000 a month. Shortly thereafter, the North Gover and Fremont Mines were opened nearby. (Courtesy Amador County Archives.)

In 1878, a newspaper correspondent visited New Chicago and reported that 30 to 40 buildings had been erected, housing three saloons, two stores, a butcher shop, a livery stable, and several boardinghouses. One of those was the New Chicago Hotel owned by Luigi (Louis) Ghiglieri. As with other hotels and saloons, Ghiglieri issued tokens to his patrons like the one pictured here. (Courtesy Amador County Archives.)

This Chinese joss house once stood in the hamlet of New Chicago. Amador County had quite a sizeable Chinese population during the Gold Rush years, with Chinatowns in Fiddletown, Jackson, and Ione. Between 1852 and 1880, the population of Chinese in California rose from 25,000 to over 300,000. After mining transitioned from surface workings to underground excavations, the number of Chinese in the gold country dwindled. (Courtesy Amador County Archives.)

Amador pioneers Frederick Charleville and Patrick Gibson established the first ranch at Bridgeport on the Cosumnes River. That land would later become the River Pines resort and eventually the town of River Pines. Charleville also owned the Union Hotel pictured in this rare 1855 view of Fiddletown's Main Street. (Courtesy Amador County Archives.)

Fiddletown had a sizeable Chinese population. In 1870, the census recorded their number at 363, most likely the historical peak. Pictured here is the Chew Kee store. Built in the early 1850s, it served as a *yao cai dian* (herb shop) with traditional medicine. The proprietor, Dr. Yee Fan-Chung, had also practiced medicine in Sacramento and Virginia City, Nevada. (Courtesy Amador County Archives.)

Dr. Yee Fan-Chung returned to China, leaving the store with his assistant Chew Kee, who also assumed responsibility for Jimmy Chow Yow, who Fan-Chung had informally adopted. When Chew Kee returned to China, he deeded the property to Yow. After Yow's death, the State of California turned the store into a museum. The artifacts that were left in the store are now on display there. (Courtesy Amador County Archives.)

When archaeologists and historians went to work cleaning up the inside of the Chew Kee store and inventorying the contents that Jimmy Chow Yow left behind, they found that he kept everything, even labels off foodstuffs such as the one pictured here. Items ranging from medicinal bottles to photographs and old calendars became part of the Chew Kee Collection. (Courtesy Amador County Archives.)

Miners arrived at the Forks of the Cosumnes, also known as Saratoga Village and later Yeomet, in 1849. The first hotel built here (above) was owned by E.P. Bowman, who also operated a ferry across the river. In 1854, he became the first postmaster, and Saratoga Village was renamed Yeomet. The settlement was situated at the preferred river crossing between Placerville and the southern mines, and by 1852, Bowman had built a suspension toll bridge across the river. The bridge can be seen in the rare image below, the only known photograph of the crossing. In 1862, Bowman sold the bridge to Samuel Huse, who later sold it to John Ballard and W.H. Martin in 1883. In 1887, William Miller bought it and was the last owner of record. (Above, courtesy Amador County Archives; below, courtesy Library of Congress.)

Four

PLYMOUTH PROSPERS

MERCHANTS AND THE COMMUNITY

John Steinbeck's *The Grapes of Wrath* and Dorothea Lange's 1936 photographic portrait of Florence Thompson titled *Destitute Pea Pickers* are modern testimonies to the hardships faced by people who migrate from their homes to unfamiliar places. History and science tell that, despite the challenges, humans have been migrating around the globe for over 200,000 years. These great movements are motivated by various needs and desires—climate, food, environmental stresses, security, and wealth, to name a few. The great Dust Bowl migration, which many alive today were witness to, was influenced by the first three factors in the list above, while the 19th-century westward migration that began with the Louisiana Purchase, followed by the Gold Rush, was fostered by the promise of wealth and land security. Those who followed the miners and early merchants to California did so along the California, Mormon, and Oregon Trails. It was "Manifest Destiny," they said—the mid-19th-century notion that the United States was destined to expand and spread democracy across North America to the West Coast and that it was ordained by God.

When Joe Williams constructed his brick trading post in 1857 at the future townsite of Plymouth, only a handful of settlers lived nearby. Most of them resided about a mile to the west at Puckerville. By 1866, a few set up housekeeping closer to Williams's business; however, the "1866 Official Map of Amador County" shows but one residence within a quarter mile—the home of Eleazer and Harriet Potter. It sat on the hillside behind the present-day location of the post office. Shortly after Potter filed his homestead, Thomas Pinder patented a large adjacent parcel to the west. These landholdings were eventually divided into town lots, and Plymouth was born. From that time to the present, the town has survived, weathering the ups and downs of the mining industry, fires, and other economic stresses. In this chapter, the reader will see the town as it evolved and meet some of the people who made it their home.

Pictured above is the earliest known photograph of the Plymouth Trading Post, also known as the Empire Store. It was built in 1857 by Joseph Chandler Williams, a native of New Hampshire who came to California around 1852. Williams first settled at Sacramento and worked for a time as a miner before going into business here. He also operated a lumberyard and stables and built a home nearby. In 1877, his home and nearly all of his holdings, with the exception of the Plymouth Trading Post, were destroyed by fire. Williams sold the brick building to the Empire Mining Company and moved to Drytown, where he opened another store that he operated until shortly before his death. Williams died on January 7, 1884, in Sacramento. He is buried in the Sacramento City Cemetery. (Above, courtesy Amador County Archives; left, authors' collection.)

The Empire Store changed facades and went through several owners over the next century. At the time the above photograph was taken around 1885, the Empire Mining Company owned the building and leased a portion of it to the Pacific Mining Company for its offices. Around 1890, the building was purchased by Hiram Edgar Potter, son of Plymouth's first settlers, Eleazer and Harriet Potter. In 1905, Bernardo Levaggi purchased the Empire Store from Potter. The building is still owned by the Levaggi family. It has been designated a California historic landmark. Hiram also served as the town's undertaker and had an office near the east end of Main Street, next to the Gazolla boardinghouse, as seen below. (Both, courtesy Amador County Archives.)

Prior to purchasing the Empire Store building from Hiram Potter in 1905, Bernardo Levaggi operated a store on the north side of Main Street, near the east end of the area now encompassed by Plymouth Park. That building, shown above, was destroyed by fire. Levaggi first came to California from his native Italy in 1863. Upon his arrival, he mined at Volcano and then opened a mercantile in West Point. In 1872, he married Mary Galliano. After he moved to Plymouth and opened a store, he also purchased the Caucasian Hall and remodeled it into a family residence. In the photograph below, Bernardo and Mary are standing behind the counter inside the Empire Store. (Both, courtesy Amador County Archives.)

Above is an 1890s view of Plymouth's Main Street. Datson's Gilt Edge Saloon, in the distance on the right, was one of six in the town at that time. Quips in Amador newspapers touted it as "the very best refreshment establishment in Plymouth." Joseph Datson, pictured at right, immigrated to America from England in 1890. After a short stint in San Francisco, he moved to Amador County and opened his saloon in Plymouth. In April 1900, Datson married Rosa Eibe Holtz. The couple's happiness was short-lived, though, as Rosa died just three months later from a miscarriage while sick with smallpox. Joseph operated his saloon until 1932, when he returned to England. The following year he married Lillian Louisa Wadge. He died in Bodmin, Cornwall, on Valentine's Day 1949. (Both, courtesy Amador County Archives.)

Sylvain Willard, a pioneer Plymouth merchant, immigrated to California from Lorraine, France, in 1851. In 1872, he entered into the mercantile business with a fellow Frenchman, Albert Falk. Their storefront on Main Street stood on the lot now occupied by the parking area for Plymouth Park. According to records in the Wells Fargo and Company Bank history room, Willard and Falk became the first agents for the company in Plymouth in 1877. Willard left the partnership, and Falk continued as agent until 1880. It is not known what happened to Willard after he left Plymouth. Albert Falk was married to Pelagie Kahn, sister of Isaac Kahn, another Plymouth merchant. After Falk sold out, he and Pelagie moved to San Francisco, where they remained until their deaths. (Both, courtesy Amador County Archives.)

Express charges do not include Duties, nor Custom House expenses, which must be guaranteed by the shipper.

Wells, Fargo & Co's Express.

READ THE CONDITIONS OF THIS RECEIPT.

Amount, $ 30

93.)

Plymouth Office, State of *Cal Jun 23* 1896

RECEIVED *from No Own Yukee*

Coin Valued at *Thirty Dols*

Addressed *No Own Yukee*

326 I St Sacrament Cal

Which we undertake to forward to the point nearest destination reached by this Company, on these conditions, namely: That WELLS, FARGO & COMPANY shall not be held liable for loss or damage, except as forwarders only, within their own lines of communication; nor for any loss or damage by fire, or casualties of navigation, and inland transportation; nor for such as can be referred to the acts of God, the restraints of Government, riot, insurrection, piracy, or the hazards of war; nor for default, neglect, or mishap on the part of any connecting or intermediate line, (individual, corporation, or association) to whom the said property may be transferred for further transmission; nor for an amount exceeding Fifty Dollars on any shipment unless its true value is herein stated. And it is further stipulated that WELLS, FARGO & COMPANY shall not be liable, under this contract, for any claim whatsoever, unless presented, in writing, within sixty days from the date hereof; and that these provisions shall extend to and inure to the benefit of each individual, corporation, or association to whom the above specified property may be transferred and entrusted in order to reach its destination.

The party accepting this Receipt thereby agrees to its conditions.

NOT NEGOTIABLE.

For the Company,

E J Bevan agt

Charges, $ 30 ¢a

MONEY RECEIPT.

Evan John Bevan, whose name appears on this receipt, became Plymouth's Wells Fargo agent in 1880. Bevan worked as a clerk in Willard and Falk's store and was a boarder in Falk's home. That same year, he was appointed Plymouth's postmaster, serving in that position until 1886. Bevan's business burned to the ground in December 1908 when a fire swept the north side of Main Street. (Courtesy Amador County Archives.)

The Plymouth Post Office has moved to various locations on Main Street over the years. This photograph was taken in 1906. Two years later, these buildings burned to the ground. Also housed here were the offices of the Butler and Talbot State Company and a telegraph office at the back of the building. A notions store occupied the building next door. (Courtesy Amador County Archives.)

John Joseph Ekel was an enigma. This attorney, who was also the first postmaster for Plymouth, had a temper that got him into trouble. In one incident, he exchanged gunfire with deputies over activities at a Plymouth house of prostitution. In 1904, he was sentenced to 18 months in San Quentin for attempting to shoot a man during a card game argument. (Courtesy Amador County Archives.)

Charles Green   Mrs Charles Green

Charles Green and his wife were not lifelong residents of Plymouth, but his ingenuity contributed a great deal to the town's economy. The couple arrived in Amador County in 1870, and two years later, he went to work for the Phoenix Mining Company in Plymouth as a mill foreman, improving the output so that more men were employed. It became the most productive mine in Amador County. (Courtesy Amador County Archives.)

Joseph Spalding Lebus was Plymouth's baker from 1912 to 1918. A native of Kentucky, he came to California via Colorado, where he married Marie Lucille Caufland in 1910. The couple had three children, Maybelle, Genella, and Orland. Two months after Orland's birth in April 1915, Marie died in Oakland. Joseph died in 1957. He is buried in the Plymouth Memorial Cemetery, next to his wife and son. (Courtesy Amador County Archives.)

Pictured here at left is the Forrest Hotel of Thomas Easton, who emigrated with his parents from England in 1831. He arrived in Plymouth in 1873, and after purchasing a lot from Eleazer Potter, Easton erected this hotel. It burned to the ground in 1877 but was rebuilt. It was again destroyed by fire in 1911, but Easton's son George chose not to rebuild. (Authors' collection.)

Plymouth residents are familiar with the Roos building on the corner of Mill and Main Streets; however, they may not be familiar with the brothers whose name is etched in stone over the entryway—Gabriel (left) and Moise (below) Roos, natives of Alsace-Lorraine, France. Gabriel came to California in 1893 and went to work for his uncle Alexander Rosenwald, who was partners with Isaac Kahn in a mercantile business on this lot. He was followed by his brother Moise in 1899. In 1911, Rosenwald and Kahn dissolved their company, and the Roos brothers took over. They operated the store through the 1940s and were also active in community organizations. Gabriel died on April 10, 1945, followed by Moise on March 3, 1949. They are buried next to each other in the Odd Fellows Lawn Cemetery in Sacramento. (Both, courtesy Amador County Archives.)

Clarence Allen Martin lived in Plymouth but a short time in the 1920s and 1930s, but he represents the many men and women from the town who have donned a uniform to serve their country. Martin may have been affected by his World War I service. After the war, he drifted around with his family in tow, not staying in one place very long. (Courtesy Amador County Archives.)

Pictured here is Plymouth physician Dr. Edward Vester Tiffany. He is associated with a legend about the tragic death of a young woman that supposedly took place in the basement of what is now the Plymouth House Inn. Truth be told, Dr. Tiffany had long since moved to Oakland when the alleged horrific crime is said to have taken place in 1921. (Courtesy Amador County Archives.)

Pictured here is the Leoni family. From left to right are (seated) Stefano Sr. and his wife, Teresia; (standing) son Herman; Herman's wife, Flora; and son Stefano. Stefano and Teresia came to California prior to 1850, settling at Diamond Springs. Their son Herman moved to Plymouth, where he and Flora were longtime residents. His Main Street blacksmith shop is featured on the front cover. (Courtesy Meredith Campbell.)

The logging and lumber business was lucrative for the Farnham family in Plymouth. This lumberyard once stood on Main Street. The pioneer patriarch of the family, Hiram Cloise Farnham, came overland to California in 1850. After residing in El Dorado County for a time, he moved to Amador and built a steam sawmill at Fiddletown. (Courtesy Amador County Archives.)

Locals will recognize the building below as Sun Ming Gee's Chinese Store, all that remains of his business, which also included a washhouse and extensive vegetable and herb garden. This was the only business in Plymouth owned by a Chinese immigrant. Gee opened the store in 1882. In 1887, he took out a mortgage on the adjacent property, where he built his laundry house and plowed a garden. Business was good, and he paid off the mortgage in 1889. Gee operated the store until his untimely death around 1916, when he was killed by a runaway team of horses. He and his wife, a mail order bride from China, had 10 children, five boys and five girls. After his death, his wife left Plymouth. Two of the daughters are pictured at right. (Right, authors' collection; below, courtesy Amador County Archives.)

The Plymouth Hotel is the only original pioneer hotel in Plymouth that escaped the many devastating fires that swept through the town. It was constructed around 1880 by William Chapman Harvey, a native of Liskeard, England. Harvey came to California in 1856, followed by his wife, Charlotte Rouse Harvey (pictured at left), in 1864. The couple had five children, Victoria Americus, California Elizabeth, Gertrude Florence, William Edward, and Joseph. The Harveys did a brisk business at the hotel, but tragedy struck when the elder William died unexpectedly in September 1883. Charlotte continued to run the hotel, which boasted a gourmet Chinese cook and a huge vegetable garden. It is said that she would feed and house up to 70 boarders at one time! In 1930, she sold the hotel to John Moreno. (Both, courtesy Amador County Archives.)

The 1930s linen-finish studio portrait at right shows the family of Juan Francisco (John Frank) Moreno and his wife, Rosa Gonzales Moreno. Their children are, from left to right, Rose, John, Ella, and Antonio "Tony." Both John Sr. and Rosa were of Spanish descent. He immigrated to the United States in 1913, and in 1930, he took over ownership of the Plymouth Hotel after the hotel he owned previously, which stood on one of the Plymouth Park lots, was destroyed by fire. John remodeled the exterior of the Harvey House to resemble the hotel that was destroyed. In the picture below, the hotel looks much different than when it was first built by William Harvey. When he died in 1934, ownership of the hotel transferred to Petar (Pete) and Luce (Lucy) Urjevich. (Both, courtesy Ella Moreno Emerson.)

The Urjevich family has long been prominent citizens in Plymouth. Martin Urjevich, pictured here in 1943, became proprietor of the Plymouth Hotel following his parent's ownership of the business. His parents, Petar and Luce, immigrated from Yugoslavia and became owners of the hotel in 1939. (Courtesy Martin Urjevich.)

The bar at the Plymouth Hotel has always been a favorite respite for locals. During the 1930s and 1940s, in addition to the partaking of adult beverages, patrons could find entertainment at the gambling tables and slot machines. This photograph was taken when the Moreno family owned the establishment. Pictured are, from left to right, hotel handyman Enrico D'Augostini, Rosa Moreno, John Moreno, and an unidentified bartender. (Courtesy Ella Emerson.)

Pictured here are multiple generations of the Ninnis family. The patriarch, Jabez Bunting Ninnis, was a pioneer settler in Plymouth. He arrived in California in 1879 and went to work in the mines. In 1883, he married Elizabeth Harvey. Jabez operated a livery stable in Plymouth for many years. He died in 1932 and is buried in the Plymouth Memorial Cemetery. (Courtesy Amador County Archives.)

Jabez Bunting Ninnis operated a livery stable out of this building. The structure was erected after the original building that stood here was destroyed in a fire in 1911. With the advent of the automobile, a livery was no longer needed, so the family converted the business into a garage. They also offered an "auto stage" service, an early name for a taxi. (Courtesy Amador County Archives.)

This busy view of Main Street was taken in 1915. The buildings on the left housed the post office and several saloons. At far right is Bernardo Levaggi's general store. Next to this stood the drugstore of A.J. Coster and then the warehouse and store of Gabriel and Moise Roos. Note that the Roos store is still a wood-frame structure. (Courtesy Amador County Archives.)

This 1920s view of the west end of Main Street shows a row of storefronts with the Levaggi home in the distance. The businesses that occupied these buildings included a confectionery, Burkes's meat market, a cobblery, a tire store, and Wheeler's grocery. The tire store may seem out of place with a wagon in the photograph, but the photograph was taken after autos came into being. (Courtesy Amador County Archives.)

A bank has occupied this Plymouth lot since about 1912 when this building was constructed. The first firm to do business here was the Bank of Amador, which merged with Wells Fargo in 1965. By 1990, it was occupied by the Bank of Lodi. There never was a Farmers Bank here; that sign was put up during the filming of a Will Rogers movie. (Authors' collection.)

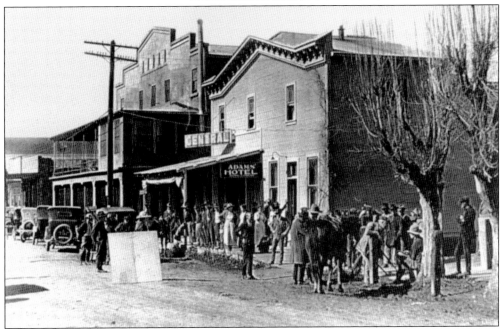

The hotels pictured here, the Plymouth Hotel (left) and the Central Rooming House (right), once stood on the lots now occupied by Plymouth Park. They were constructed after a fire in 1908 that destroyed the buildings previously occupying these lots. In 1931, another fire burned the north side of Main Street, and both of these hotels were lost. (Authors' collection.)

Methodist Episcopal services were first held in this church, which now serves as the gathering place for Drytown Masonic Lodge No. 174. In 1882, the church purchased the lot from Thomas Pinder and erected the church. The lodge purchased the property and building from the church in 1926. Since its inception, the Drytown Masonic Lodge experienced many trials. Tyro Lodge No. 73 was formed in Drytown in 1855 but dissolved four years later. Ten years later, former members of Tyro were granted a charter to establish Lodge No. 174. When their old building in Drytown began to decay, members moved their meetings to the old Plymouth Odd Fellows Hall on Main Street. After that building burned to the ground, the Masons purchased the Methodist Episcopal church and have held their meetings here ever since. (Courtesy Amador County Archives.)

St. Mary's of the Mountain served as the worship center for Plymouth's Catholics until August 2020 when it was closed and subsequently sold into private ownership. The first Catholic church constructed on this site was built in 1881 under the pastorate of Fr. Rafael De Cardis. He served this community until his departure for Italy in 1886, where he died in 1934. The church that now stands here was built in 1939 under Fr. Joseph Hannrahan, who ministered here from 1930 to 1949. The building was designed by C.C. Cuff of Sacramento, who also presented the marble altar to the congregation. Irene Levaggi Chichizola donated the oak pews in memory of her parents, Bernardo and Mary Levaggi. A life-sized scene of the crucifixion was painted over the main altar by Leona Garibaldi. The new St. Mary's was dedicated on April 23, 1939, by Bishop Robert Armstrong of the Sacramento Diocese. Fr. Colin Wen of St. Katherine Drexel Parish led the final mass held here on August 15, 2020. (Authors' collection.)

Plymouth residents have long been inclusive of those of different faiths. Catholics, Christians, and Jews have lived and worshipped here since the first homes were built. In the above photograph, Rev. Joe Mays of Drytown (left) and congregants gather for the 1964 groundbreaking for the First Baptist Church of Plymouth. The Baptist mission was started in March 1962 and was organized into an independent church in March 1964. Below, from left to right, Pastor Linda Lange and Fiddletown residents Marie and Ron Scofield stand in front of the Plymouth Community Church. This church performs a lot of community outreach, including its Help for the Holidays program and feeding the needy with its mostly organic food closet. (Above, courtesy Amador County Archives; below, courtesy Michael Spinetta.)

The Plymouth Memorial Cemetery covers ground originally taken up under two private land patents. The west half is on land that Thomas Pinder filed a homestead for in 1865. His donation to the city for a public cemetery was made in May 1876. The eastern half sits on land that was donated to the city shortly after it was taken up under an agricultural patent by Eleazer Potter in 1879. It has been in continuous use since the 1870s, and many of the town's pioneers are buried here. The photograph at right shows a memorial to miners who were lost in a disastrous fire at the Fremont Mine east of Drytown. Shortly after lunch on November 30, 1907, thirteen miners were lowered into the Fremont shaft. Unbeknownst to those above ground, a fire had ignited in the mine. Just below the 1,000-foot level, the men encountered smoke. Confusion ensued and only two men were able to escape to the ladder and make their way to the surface. The other 11 perished, 10 of whom were residents of Plymouth. (Both, authors' collection.)

The students posed above could very well be the children of those in the photograph below. They were taken about 35 to 40 years apart at the old Plymouth school that once stood on the hill behind the Plymouth Park lots. The above image was taken in the 1920s and the one below in the 1890s. Children in this area first attended classes in a one-room schoolhouse at Puckerville, about one-half mile west of Plymouth proper. As the population grew, a new schoolhouse was built. When that school was burned to the ground by an arsonist, this building replaced it. The bell that was housed in the steeple of this building is now mounted on a pedestal at the Plymouth Elementary School. (Both, courtesy Amador County Archives.)

Five generations of the Colburn family have lived in and served the community of Plymouth. Pharmacist Harold Colburn (above, fifth from left) and wife Marie (seventh from left) operated the Plymouth Drugstore for many years and raised their five children, Marla, Nan, Harold Jr., Gary, and Jon, in their two-story Victorian home on Main Street. Harold Sr. was also active in the community as a county supervisor, a school board member, a city councilman, mayor of Plymouth, the first chairman of the Amador County Fair, a fair board member for 44 years, and a lieutenant governor candidate. Below, four generations of Colburns stand around the founding fathers of Amador County Fair plaque in 2018. (Above, courtesy Amador County Archives; below, courtesy Amador Ledger Dispatch.)

The Native Sons and Native Daughters of the Golden West have been a presence in Plymouth since the late 19th century. Native Sons Parlor No. 48 was founded in 1885. It held its first meetings in the Independent Order of Odd Fellows hall on Main Street, and in 1915, the Native Sons built a hall on Mill Street. That building is now a private residence. The Native Daughters Forrest Parlor No. 86 was established in 1895. Above, Parlor No. 48 celebrates its 75th anniversary in 1960. From left to right are state president Leonard King, parlor president Frank Dal Porto, and grand trustee Frank Chrisi. Below, Hazel Marre (second row, second from left), Marion Vaira (second row, third from left), and other Amador members of the Native Daughters Forrest Parlor gather for a visit of the grand officers to Amador County in 1956. (Both, courtesy Amador County Archives.)

Five

MAKING MERRIMENT
SPORTS, CAVALCADES, AND A FAIR

As the famed 15th-century English bishop Joseph Hall wrote, "Recreation is intended to the mind as whetting is to the scythe, to sharpen the edge of it, which otherwise would grow dull and blunt." In other words, life is dull without fun.

Since the Gold Rush era, there has never been a shortage of entertainment and recreational opportunities for the residents of northwestern Amador County. Parties, potlucks, parades, dances, a movie, the circus, gambling, a fair, live music, a play, hiking, fishing, swimming—all the wonderful things that carry people away from the doldrums of everyday life—have long been traditions in this neck of the woods. The miners, the first Euro-Americans to set foot in this region, spent hours gambling and playing mumbley-peg. As the camps became larger, traveling troupes of actors and circus performers were the first to formally entertain here. Some towns, like Volcano and Drytown, actually had local thespian groups. By the late 19th century, nearly every town had a brass band or a string quartet that provided music for gatherings and dances, not to mention the "porch" musicians who played fiddle, guitar, banjo, and other instruments at birthdays and anniversaries.

The biggest recreational event of the year in Plymouth is the Amador County Fair. For many decades, the 26th Agricultural District Fair was held alternately in Sacramento and Amador Counties, both of which made up the 26th District. Then in 1938, after Sacramento and Amador were split from the district, the Plymouth fairgrounds opened and became the permanent home of the Amador County Fair. In 1939, a special exhibit featuring Amador's mining industry included an operating model stamp mill with a miniature headframe and ore samples. In Fiddletown, the annual Fiddler's Jam draws crowds of music enthusiasts from near and far. And of course, there are the local pubs where folks gather to hear live music and socialize, and clubs such as the Boy and Girl Scouts and 4-H for children. Plymouth annually hosts a Fourth of July celebration, while the 49er RV Park offers children and adults alike spooktacular fun at Halloween.

Nearly every small town in Amador had a brass band to entertain at functions ranging from parades to political rallies. The first Plymouth Brass Band, pictured here, was organized in 1897. Alas, the names of its members are not known. The tent behind them was home to the local photographer who set up shop on the lot behind Caucasian Hall. (Courtesy Amador County Archives.)

This photograph was taken during the Plymouth Fourth of July parade in 1908. The wagon in the lead and the women with parasols were entered by Plymouth's Native Daughters of the Golden West Forrest Parlor No. 86. On the left, decorated with flags and banners, is the Roos store. Note the cover over the porch. This would burn in a fire that struck later that year. (Courtesy Amador County Archives.)

Benevolent organizations have long been a means by which members can fraternize with fellow citizens and celebrate with the community. Pictured here are young members of the Native Daughters of the Golden West Forrest Parlor No. 86 in preparation for a Flag Day celebration. The photograph was taken around 1900. (Courtesy Amador County Archives.)

Entertainment comes in all forms. For many of the rural folks in northwestern Amador County, this included recreation on the farm, like a leisurely horseback ride across the countryside with a friend. This photograph of two unidentified young women was taken on one of the farms a short distance from Plymouth. (Courtesy Amador County Archives.)

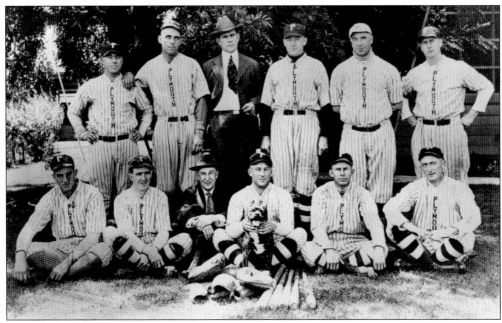

Ah, the great American pastime—baseball, or as it was once known, "Base Ball." Every town in Amador County boasted a team, and Ione even had a Native American one. Rivalry was rampant, with each town cheering their hometown athletes and placing bets on the outcome of each game. The teams pictured here graced the diamonds in the late 19th century. Today, the sport is kept alive with T-ball, Little League, junior and high school teams, and the Gold Country Vintage Base Ball League. The latter claims to be more than just old-fashioned uniforms; it plays baseball "the way the game was meant to be played"—with spirit, sportsmanship, etiquette, and a little of the "outlandishness" from the early days. (Both, courtesy Amador County Archives.)

California, with its rich agricultural industry, is known as the "Breadbasket of the World," so it is no surprise that citizens take pride in their agricultural county fairs. Amador and Sacramento Counties once comprised California's 26th Agricultural District, and their fairs were held alternately between the two places. Ione first hosted it in 1887. Eventually, the district was split in two, and the fair came to Plymouth in 1938. That first fair was held in a large tent on land that is now occupied by the Plymouth Elementary School playground, which was eventually purchased by the fair. The photograph above shows a geologic display of specimens from Amador County at the 1889 26th Agricultural District Fair held in Sacramento. Below is the tent in which the 1938 Amador County Fair was held. (Both, courtesy Amador County Archives.)

Between 1938 and 1955, nearly 40 acres of property was purchased from various landowners for the Plymouth fairgrounds, including J.M and Gladys Loyd, the Lubenko family, the W.F. Detert estate, and the Grielich family. Part of the funding came from a Fairs and Exposition fund, which received a percentage of the money bet on horse racing in California. A portion of land, 5.5 acres, was sold back to the school district for Plymouth Elementary School. The fair was charged a nominal cost for the 23 acres acquired from the Detert estate. As seen in the photograph at left, in appreciation of the Deterts' generosity, a plaque was placed on the flagpole base at the fair in memory of W.F. Detert. Below is the entrance to the newly established fairgrounds. (Both, courtesy Amador County Archives.)

What is a fair without celebrities? In 1941, former world heavyweight boxing champion Max Baer participated in boxing exhibitions at the Amador County Fair. Baer held the world heavyweight title from June 1934 to June 1935. He fought a total of 81 fights over his boxing career, with 68 wins and only 13 losses. Of his wins, 51 were by knockout. (Courtesy Amador County Archives.)

The Miss Amador Pageant is an annual event with the queen being crowned at the Amador County Fair. The pageant contestants in this 1964 photograph are, from left to right, Kathy Goulding, Sharon Ayres, Bernadette Sowerby and Luaina Mattson (on ladder), Penny Nichols, Sharon Ingersol, Sharen Sheridan, Bonnie Bonham, Nancy Gratt and Gail Leake (on ladder), Penny Ann Faddis, and unidentified. (Courtesy Amador County Archives.)

Livestock is the main attraction of agricultural fairs. From poultry and rabbits to cattle, pigs, sheep, and goats, Amador County Fair participants take pride in the animals they show. A favorite are the horses and the rodeo, the latter held on Friday evening. In this 1950s photograph, fair manager Goula Wait presents a horsewoman with the Grand Champion award. (Courtesy Amador County Archives.)

One of the hubs of the Amador County Fair is Old Pokerville. This row of buildings first opened at the fair in 1962 and was dedicated by Gov. Edmund G. Brown. It is on the green in front of the buildings that old west and mountain men reenactments are held to entertain visitors. This photograph was taken in the 1950s. (Courtesy Amador County Archives.)

Ephemera collectors look forward to the Amador County Fair poster illustrated by Amador's best loved artist, Rand Huggett. For decades, Huggett has taken on this task, creating artwork in line with the theme chosen for that year. In 2021, "Back in the Saddle Again" was appropriately chosen, given the 2020 fair was limited to livestock judging due to the COVID-19 pandemic. (Courtesy Amador County Fair Board.)

A favorite event at the Amador County Fair is the rodeo. The event opens with the singing of the national anthem, while a rider on horseback circles the arena with the flag held aloft. Competitions include roping, bronc riding, bull riding, and mutton busting. In this photograph, taken at the 2021 fair, Seth Seaver (holding microphone) pays tribute to former fair officials before the rodeo gets underway. (Courtesy *Amador Ledger Dispatch*.)

The California Department of Forestry and Fire Protection takes the opportunity every year at the fair to educate citizens about the dangers of fire. One way is to entertain children with its mascot, Captain Cal. In this 2021 picture taken at the fair are, from left to right, Susan, Connor, and Sean Coffey, who pose with their newfound friend. (Authors' collection.)

In addition to the standard amusements, the Amador County Fair offers visitors top-notch entertainment on three stages at the fairgrounds. Bands, magic shows, comedians, and a talent contest are all part of the annual event. In this 2015 photograph, the swing band Mr. Pinstripe performs on the Biagi Stage. Members of the band, which is no longer together, were all Amador County residents. (Authors' collection.)

In the above photograph, the pit crew from Campbell's Construction Company make quick repairs to their car between heats of the Amador County Fair Demolition Derby. The derby, held every year on Saturday evening, is one of the highlights of the fair and a fairgoer favorite. The bleachers of the rodeo grounds are filled to capacity, with standing room only along the fences. The roar of the engines, the smashing of metal, and the cheers of the fans can be heard beyond the fairgrounds for over a mile. In the photograph below, safety takes priority as drivers of disabled cars are required to stay in their vehicles and wave a white flag to notify active vehicles that they are now out of the competition. (Both, courtesy Michael Spinetta.)

This photograph, titled "Plymouth Thespians" by its original owner, was taken around 1930. The actors apparently entertained their fellow citizens. From left to right are (first row) Jessie Mae Laberdie and unidentified; (second row) unidentified, Mildred Landrum, and Gladys and John Moyle, who ran the Plymouth Garage. (Courtesy Amador County Archives.)

Plymouth's movie theater, at far right, entertained film fans during the mid-20th century. The ticket booth that saw many a long line and the screen that delighted adults and children alike for decades are long gone. After the theater closed, the building was turned into apartments. Several years ago, it was again remodeled to become a hotel known as the Rest. (Courtesy Amador County Archives.)

How Fiddletown got its name is a bit of a mystery. Tax assessor's records from 1852 refer to a Fiddlers Flat, while Jesse Mason's 1882 history of Amador County claims it was named after a group of Missouri miners who were always fiddling around. In any case, the name stuck—except between 1878 and 1932, when it was known as Oleta! What better symbol to represent the town than the larger-than-life fiddle that sits atop the community hall. The fiddle was rededicated in 2021 at the 70th annual Fiddlers' Jam, an event that has been held since 1951. Everybody loves a parade and that includes the residents of Fiddletown, as can be seen in the photograph below. (Both, courtesy Amador County Archives.)

Fiddletown's first Fiddlers' Jam was held in July 1951. A visit by the famed Russian violinist David Rubinoff was the highlight of the event. Rubinoff judged the fiddler's contest, selecting Rudolf Schlein of Garden Valley, California, as the winner. Afterwards, the virtuoso entertained the crowd with several selections played on his Stradivarius. (Courtesy Amador County Archives.)

The Fiddlers' Jam draws talented players of many stringed instruments from all over California. Visitors to this small festival are entertained with stage shows featuring bluegrass bands, jam sessions, and impromptu open mic singers and players. A country barbecue and vendor booths are also part of the day's festivities. This photograph was taken at the 2021 Fiddlers' Jam. (Courtesy Michael Spinetta.)

The festive holiday season is always a reason for entertainment. In the photograph above, Amber Spinetta and Ron Scofield practice tunes they will play at the Fiddletown Carols and Cookies celebration held at the Fiddletown Community Center in December 2021. Scofield and his wife, Marie, are owners of the Red Mule Ranch in Fiddletown, where they host the annual Cowboy Campfire. Well-known entertainers in the western music genre perform while visitors are treated to an evening of cowboy barbecue and music. Ron is also a talented artist and wheelwright. His gift for restoring stagecoaches, buggies, and wagons is known across the country. He is a regular at the Amador County Fair, where he demonstrates the art of his craft. The logo for the Cowboy Campfire (below) was drawn by Ron. (Above, courtesy Michael Spinetta; below, courtesy Ron Scofield.)

Fiddle music was not restricted to just Fiddletown. In this photograph, members of the Matulich family gather with trumpet, guitar, and fiddle in hand for a New Year's celebration in 1908. From left to right are brothers George, John, and Alexander "Alec" Matulich. This photograph was taken in front of the milk house at their home in Drytown. (Courtesy Amador County Archives.)

In addition to community events, many entertainments were sponsored by local churches. In this 1891 photograph, a Sunday school picnic is held on the Fourth of July. Note the arch standing on a small stage beneath the trees at left, where the minister presented his sermon. Off to the right, under another stand of trees, women hand out food and beverages. (Courtesy Amador County Archives.)

Agriculture plays an important role in the lives of northwestern Amador County residents. Traditionally, harvest festivals were held at the beginning of harvest season; however, given the grape harvest, the Plymouth Foothills Rotary Club hosts its Harvest Fest at the end of October. Games, music, good food, and wine tasting at local businesses are all part of the event. (Courtesy Plymouth Foothills Rotary Club.)

Residents in this neck of the woods go all out to make sure their children have a fun-filled Halloween. The 49er RV Park on State Route 49 is a favorite place to gather and celebrate the holiday. Children wind their way through spooky decorations while trick or treating. Pictured from left to right at the 2021 Halloween party are Hannah Lee, Penny Simmons, Susan Coffey, Sean Coffey, and Connor Coffey. (Authors' collection.)

97

The tradition of musical entertainment has been kept alive at the Drytown Club. Known as the "only wet spot in Drytown," it has long been the gathering place for locals and out-of-town visitors. Here, cold beer, wine, and soft drinks have been readily available since the mid-20th century. In the photograph above are Laura Doyle (left) and Allen Frank (right), former owners of the tavern at different times. Here, they play a last hurrah at the closing of the club in April 2021. Allen's Doghouse Blues Band drew fans from as far away as the San Francisco Bay Area. On drums is Allen Kreutzer, owner of the neighboring Drytown Cellars. The building is now home to Feist Wines Drytown Social Club. (Above, courtesy Eric Marks; below, authors' collection.)

Six

DID SOMEONE YELL FIRE?

DEVASTATION IN A SMALL TOWN

Fire was the bane of every 19th-century American town. Wood-framed structures could be consumed by flames in the blink of an eye. Fires were often ignited by poorly ventilated or unattended wood stoves, while others were acts of nature that began in the wildlands and spread to consume entire towns. And then there was arson. As mining camps grew into towns, residents were called on to extinguish the blazes by dumping buckets of water on the flames, covering them with dirt, beating them with gunny sacks, and sometimes stomping on the fire. They often placed barrels underneath downspouts to catch rain or would fill them with water from wells or nearby watercourses to be ready should a blaze erupt. Eventually, volunteer fire departments were established, and the latest in firefighting equipment stood at the ready. Insurance companies entered the picture as towns expanded. The underwriters required that certain safety measures be put into place to avoid the total destruction of buildings. One method was to fill the space between the ceilings and floors in multistory structures with soil or sand. The idea was that as the building burned, the soil would drop to the lower level and extinguish the flames.

The towns in northwestern Amador County were no exception when it came to destruction from fires. Plymouth experienced five major fires during its history. These occurred in 1874, 1877, 1908, 1911, and 1931. Each of these conflagrations destroyed numerous buildings. Drytown was also plagued with disastrous infernos, the most devastating in 1857. Every little hamlet suffered losses. Many of the burned structures were rebuilt with brick, but even those could succumb to destruction if their interior framing was consumed by flames.

Today, many small-town fire departments are still volunteer. In Amador County, Jackson, Ione, and Volcano rely on volunteers. Fires stations are now located in all Amador towns, a number of them under the jurisdiction of the Amador Fire Protection District like the one in Plymouth. These departments are assisted in their firefighting efforts by the California Department of Forestry and Fire Protection.

This fire bell tower has stood on the hill behind city hall for several generations. It was first erected on Main Street when it began its job of sending out an alarm to the town at the onset of a blaze. The bell was also used during World War II to alert citizens that a civil defense blackout was being activated. (Authors' collection.)

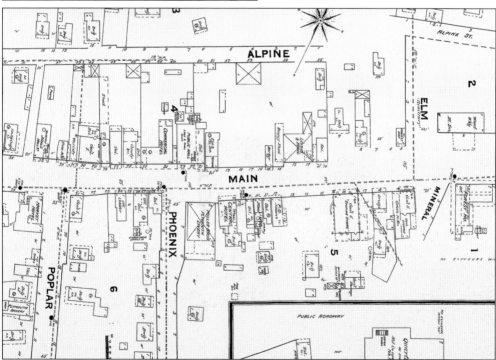

Plymouth, in its infancy, was dependent on water stored in barrels around town and water in the Plymouth Ditch, if any, to extinguish fires. By 1895, when this map was drawn by the Sanborn Fire Insurance Company, the Empire Mining Company had installed pipes and hydrants along Main Street. The hydrants are the dark dots on the map. (Authors' collection.)

The row of buildings shown here were built on the Plymouth Park lots after the 1877 fire. From left to right are the Adams Hotel, also known as the Commercial Hotel; the town hall, also known as Norris Hall, which housed a dance floor, a saloon, and a barber; Thom's variety store; and the telephone exchange and post office building. (Courtesy Amador County Archives.)

This photograph of the north side of Plymouth's Main Street shows the aftermath of the 1908 fire. It started in the flue of the Adams Hotel, which stood where the parking area for Plymouth Park is now located. Nine businesses and one home were destroyed, including Lawrence Burke's first butcher shop. (Courtesy Amador County Archives.)

In 1911, fire once again visited Plymouth. That blaze started in the laundry room of George Easton's Forrest Hotel, which stood at the east end of Main Street. The Plymouth Ditch, which provided the town with water, was being cleaned out, and thus the town was left without a good supply to fight the fire. The Central Hotel, pictured here, was also destroyed in the blaze. (Courtesy Amador County Archives.)

The Plymouth firehouse once stood on the lot to the east of the space now occupied by the parking area for Plymouth Park. In 1931, a fire broke out in the Adams Hotel and spread along the north side of Main Street. The fire department building, hotel, city hall, Alpine Garage, and a vacant store were completely destroyed. (Courtesy Amador County Archives.)

The Plymouth firehouse is now located on Sherwood Street across from the Amador County fairgrounds. As part of the Amador Fire Protection District, Station No. 122 has a year-round paid staff assisted by volunteers and administered by Chief Walt W. White. Housed here are engines 5221 and 5222 and water tender 5223. (Courtesy Michael Spinetta.)

The Fiddletown Schoolhouse has stood at this site since 1862. It replaced an earlier building that was constructed in the 1850s. In January 1933, fire threatened to destroy this structure when flames erupted from the roof. Thankfully, volunteers were able to extinguish the blaze using snow on the ground from a recent storm. (Courtesy Amador County Archives.)

The first fire hydrants were installed in Plymouth by the Plymouth Consolidated Mining Company. Here, residents look on as firefighters check the flow of a hydrant. Fire departments must be prepared for the extremely dry conditions that occur annually in California. In recent years, drought conditions have led to conflagrations that are much larger in scope than any in the state's history. (Courtesy Amador County Archives.)

Pictured here are teachers and students at the old Drytown School. This school, probably the first built in the town, caught fire and nearly burned to the ground. In 1883, the school district superintendent visited the school and counted 76 students between the ages of 5 and 17. (Courtesy Amador County Archives.)

Seven

BOOM AND BUST
THE RISE AND FALL OF THE MINES

When miners first flocked to the Sierra Nevada foothills, they dreamed of rivers and streams filled with gold nuggets waiting to be plucked from the water. When they arrived, they found it to be much different than they had imagined. They soon learned that the precious metal did not give itself up easily. It took day after day of backbreaking work. Although some did make small fortunes during the early days of the Gold Rush, it was not until the late 19th and early 20th century, when hard rock mines opened, that the real fortunes were made.

In northwestern Amador County, the first claims were worked along the Cosumnes River and streams that cascaded from the Sierra. Later, claims for subsurface workings were filed throughout the foothills along the mother lode belt. Near the future townsite of Plymouth, the Aden and Simpson claim was the first to be mined. It eventually become the Hooper's Plymouth Quartz Mine, for which the town is named. The most well-known mine in this area was the Plymouth Consolidated, a conglomerate formed from the combined Phoenix, Hooper, Oaks, and Pacific claims. Money poured into the operation from San Francisco investors, and it proved to be profitable. However, it was not until an Italian prince showed up that production skyrocketed to make it the richest producing mine in California. Elsewhere, near Drytown, the Pioneer, Potosi, Drytown Consolidated, and other mines turned the once small camp into a boomtown. At New Chicago, the Fremont and Gover produced large profits as well. Surrounding Fiddletown, gold placers and marble quarries were worked.

Investors made money hand over fist while miners worked for meager wages until unions were formed. Accidents were common in the mines. In 1907, a fire broke out in the Fremont Consolidated that claimed the lives of 11 miners. Another fire at the Plymouth Consolidated burned for six months and forced its closure for over two decades. During World War II, all of the large mines in the United States were closed down under presidential order. Many were never worked again.

John Hooper was a member of the Hooper family of San Francisco, who made their fortune in the lumber business. He purchased an interest in the Simpson and Aden Mine, which was the first claim filed at the future townsite of Plymouth. Within a short time, he was sole owner and renamed the claim Hooper's Plymouth Quartz Mine, for which the town is named. (Authors' collection.)

The Alpine Mine, later known as the Wheeler Mine, was another early claim in Plymouth. It sat on the hill just east of the cemetery. This claim operated off and on between the 1850s and 1942, when the mines were closed down during World War II. The concrete footings for the 10-stamp mill and several small buildings remain on the property. (Courtesy Amador County Archives.)

The Amador Star Mine, also worked under the names Hamberger Mine, Rhetta Mine, and West American Consolidated Gold Mines, was three miles north of Plymouth. It was first opened up in the 1870s but very little work was done here. Work was sporadic; its most successful period was from 1932 to 1935, when over $20,000 worth of gold was taken out. (Courtesy Amador County Archives.)

The Bay State Mine, later worked as the Venture Quartz Mine, was located about four miles north of Plymouth, near the settlement of Enterprise. It was first opened up in the early 1890s and produced about $34,000 in gold between 1892 and 1897. After this, it was worked off and on until 1935. For a time, it was operated in conjunction with the Amador Star Mine. (Courtesy Amador County Archives.)

The Pioneer Quartz Mine was south of Plymouth proper. This mining claim was filed in 1852. The workings here include an inclined shaft that extends 550 feet deep. Gold and silver were extracted at five levels, occurring at 100-foot intervals. Over $40,000 in ore was produced here between 1895 and 1949. Small amounts of lead and copper were also produced at the Pioneer. (Courtesy Amador County Archives.)

Pictured here is an inside view of the stamp mill at the Pioneer Quartz Mine showing a series of shaking tables. Nearly all of the larger mines had a mill on-site to process ore. The man standing on the table at left is unidentified. Standing between the tables is Volcano druggist Dr. David Boydston, who provided potassium cyanide and sulfuric acid to mine operators. (Courtesy Amador County Archives.)

This photograph of the Plymouth Consolidated Mine headframe was taken in 1940 when the mine was undergoing renovation. This was first filed as the Simpson and Aden claim and was later purchased by John Hooper. In 1871, Hooper sold it to California mining magnate Alvinza Hayward, who renamed it the Empire. In 1883, it became part of the Plymouth Consolidated. (Courtesy Amador County Archives.)

Pictured here are mine shaft foreman Chezro Martin (left) and mining engineer Hamilton Lewers underground in the Empire shaft of the Plymouth Consolidated Mine. Martin worked nearly his entire life in the mines, beginning as a miner, then as a powder man, and eventually a foreman. He also served on the Plymouth City Council and was active in public affairs. (Courtesy Amador County Archives.)

Gelasio Caetani was an Italian nobleman. He graduated from the Columbia University School of Mines in 1903 and after mining in Idaho, came to California and was hired by the Plymouth Consolidated Mining Company to work as an engineer. His invention of a new mill conveyor system that was installed here improved production, making the mine the largest producer in California for decades. (Courtesy California Division of Mines.)

The Plymouth Consolidated Mining Company was a conglomerate that came into existence in 1883. Its holdings, which covered 126.3 acres, included the Empire, Pacific, Plymouth Quartz, Southerland, Oaks, Pacific, Simpson and Aden, Reese and Woolford, Phoenix East, Indiana, Rising Star, Conville, and Beta claims, as well as the Phoenix Mill site. The Plymouth Consolidated was closed by presidential order in 1942. (Courtesy Amador County Archives.)

The Ballard Mine, another claim north of Plymouth near Enterprise, was first worked in the 1870s but then closed down until 1928 when it was mined briefly. The largest amount of work here was conducted during the 1930s and 1940s, but like other mines, it was shut down in 1942. After World War II, a small amount of development took place in 1947. (Courtesy Amador County Archives.)

Mechanized dredging along waterways in California began in 1898. Extracting gold from river gravels, which was first done by miners with pans and sluices, took place on a larger scale by employing floating bucket-line dredges like the one pictured here. This one, known as the Placeritas Dredge, operated on the Cosumnes River near Plymouth. (Courtesy Amador County Archives.)

This photograph was taken underground at the Gover Mine, east of Drytown. Working conditions for miners were horrific in the early days. In the early 1860s, the hard rock miner's union movement began in Storey County, Nevada, and soon spread to California, where over time, conditions and wages improved. Note the woman at right. (Courtesy Amador County Archives.)

This picture of the Engineers Mine near Fiddletown, then known as Oleta, was taken in 1922. It was owned by the Engineers Mining and Leasing Syndicate. This mine operated sporadically under a multitude of names from the 1880s through the 1940s and only produced small amounts of gold. Like other mines, it was closed down in 1942 by order of the US War Production Board. (Courtesy Amador County Archives.)

Eight

AGRICULTURAL PURSUITS
HAY, LIVESTOCK, AND GRAPES

In 1849, at the onset of the Gold Rush, most who headed to California did so with the intent to get rich mining, but others saw an opportunity to build wealth supplying goods and services. It was not long before mercantiles, taverns, clothiers, haberdashers, hotels, laundries, hardware stores, and other businesses established a lucrative trade. Like the goods that stocked the stores during the first few years of the Gold Rush, much of the agricultural products consumed in the goldfields were brought in from San Francisco, Sacramento, and Stockton, with the exception of a small portion produced in local vegetable gardens and newly established livestock ranches.

Although it is generally believed that the Gold Rush economy was sustained by the mining industry, by 1850 it began to diversify. With large tracts of land available, farms and ranches were soon established and began to produce goods. Within 10 years, about 36,000 acres were under cultivation in Amador County and livestock was valued at over $750,000. By 1860, an average-size farm in the county was 395 acres. Nearly everything from cattle, sheep, and goats to market produce, nuts, grain, and grapes were being produced and processed locally.

Butchers processed meat, while gristmills at Sloughhouse, Plymouth, and Ione readied grains for both human consumption and livestock feed. Plymouth's first small-grain mill was located on Main Street. After it burned to the ground, another larger mill was built north of town on the road to Placerville. Every town had one or more butchers and local dry goods stores, which sold both fresh produce and grain. John Daveggio had the first large truck garden in Plymouth on the south side of Main Street. It covered several acres between Main and Mineral Streets. Near the west end of Main Street, Charlotte Harvey had a large garden of produce to feed patrons at the Harvey House. To the east in the Shenandoah Valley, Swiss immigrant Adam Uhlinger established the first winery in the region, the beginning of an industry that is world-renowned today.

Berthal Leslie Crain was a native of the Shenandoah Valley, born there on October 7, 1896, to Jackson and Olive Ball Crain. Known as "Ted" to his family and friends, he was a lifelong farmer. Crain served his country in Europe during World War I and was an active member of the Native Sons of the Golden West and the Shenandoah Valley–Willow Springs Grange. (Courtesy Amador County Archives.)

Hay farming was one of the first agricultural industries developed in northwestern Amador County. The earliest hay farms were near the Cosumnes River, where water was plentiful and could be diverted to fields during the dry season through irrigation ditches. In this photograph, John Hass drives the lead wagon, hauling hay to be stored in his barn. (Courtesy Amador County Archives.)

Pictured here seated around a hay bailer are, from left to right, Rosalie Estey, Don Estey, Earl Estey, Fred Estey, Hugh "Henry" Bell, and Clyde Estey. Both the Estey and Bell families were early settlers in the Shenandoah Valley who took up agriculture for their livelihood. Descendants of both families still reside in northwestern Amador County. (Courtesy Amador County Archives.)

Agriculture, on both a large and small scale, is still a vital part of the economy in this region of Amador County. In Plymouth, the Plymouth Community Garden provides space for residents to grow both vegetables and flowers. Situated at the foot of the hill below the Independent Order of Odd Fellows lodge building, the garden also serves as a source of educating the community about gardening. (Courtesy Michael Spinetta.)

The equine industry has been a vital part of the economy here since the early days of the Gold Rush when dray horses and mules were used to transport goods by wagon. Mules were also used to work arrastras to crush ore. The horse show and rodeo have been a part of the Amador County Fair since its beginnings and are a favorite of fairgoers. (Authors' collection.)

Nationwide, nearly six million children participate in 4-H every year, learning through hands-on projects in the areas of agriculture, science, health, civic service, and home arts. The annual Amador County Fair provides youth with a venue to show their accomplishments, including their farm animals. In this photograph, from left to right, 4-H members Tennessee and Lucas Burns and Amber Spinetta pose for a photograph at the fairgrounds. (Courtesy Michael Spinetta.)

In this 1961 photograph, the plaque marking D'Agostini Winery as California Historic Landmark No. 762 is unveiled. This winery, one of the oldest in California, was founded in 1856 by Swiss immigrant Adam Uhlinger. In 1911, it was purchased by Enrich D'Agostini. It is currently owned by the Sobon family. Some of the original vines planted by Uhlinger are still in production. (Courtesy Amador County Archives.)

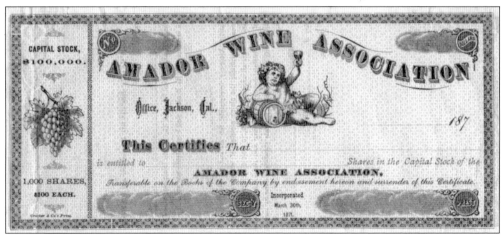

Within two decades of the Gold Rush, a number of wineries were established in the Shenandoah Valley and in the hills around Fiddletown. As with any industry, cooperation is a necessity to promote common endeavors. The first Amador Wine Association was established by 1871, supposedly to improve the production and sale of grapes and wine. It turned out to be a failed subterfuge. (Courtesy Amador County Archives.)

Charles Spinetta, grandson of Italian immigrants Giovanni and Natalina Spinetta, continues the family farming tradition that began in Amador County in 1852. He is seen here in the tasting room of the Spinetta Winery and Wildlife Art Gallery. Spinetta offers a variety of wines, including Barbera, Chenin Blanc, Petite Sirah, Primitivo, three styles of Zinfandel, and rare aged Amador County wines. (Courtesy Amador County Archives.)

The Deaver Vineyards tasting room is a favorite stop for both locals and visitors touring Shenandoah Valley wineries. Ken Deaver Jr. and his wife, Jeanne, operate the tasting room in conjunction with vineyards worked by his father before him and his step-great-grandfather John Davis, who first planted Mission and Zinfandel grapes in the 1860s. Ken opened Deaver Vineyard on Steiner Road in 1985 and also owns the Amador Flower Farm. (Authors' collection.)

Nine

NEIGHBORS NEAR AND FAR

FROM DRYTOWN TO THE COUNTY LINES

When immigrants first traveled to California, many came alone while others moved west in groups on wagon trains. The former were mostly miners and adventurous tradesmen, while the latter consisted of family groups and neighbors. The tradition of extended families relocating in America began when this country's first immigrants moved beyond the boundaries of the established colonies. This tradition continued into the 19th century and beyond.

In 1849, Missourian Andrew H. Hinkson captained a wagon train that brought the Bell, Boone, Hinkson, and other families to California. Andrew Jackson Crain, who would settle in the Shenandoah Valley, served as a scout on the westward journey. Hinkson and the Boones settled in Drytown, where Thomas Silvera from Portugal became a butcher, E.W. Watton was the blacksmith, and John Braun from Germany served as the town barber. James Wheeler, another settler in the Shenandoah Valley in the early 1850s, was the first miller in the region. By the mid-1850s, Fiddletown was also settled. Native American families like the Howdy clan lived and worked alongside their neighbors—the Clarks, Isaac Cooper and family (nephew of James Fenimore Cooper), the Farnhams, the Votaws, and others. A substantial Chinese population occupied a section of the town, with businesses that included a gambling hall and the Chew Kee Store, among others. Plymouth, still a sparsely populated settlement until about 1870, became home to the Potter, Easton, Harvey, Estey, Ninnis Wheeler, and Miller families, to name just a few. Ranchland surrounding the town was taken up by men such as Joseph Woolford, inventor of the reversible ratchet wrench, and William Kroning, who at the young age of 23 was elected justice of the peace for the township.

It was the cooperation between these neighbors that helped to turn fledgling settlements into towns, establish successful businesses, and instill a sense of community. Oh, that this tome was much larger to include the many people who made the northwestern region of Amador County what it is today. Alas, readers must settle for the limited space here and meet but a few of those neighbors.

Joseph Davis, a lifelong resident of the Shenandoah Valley, was born there in 1873 to pioneers John James Davis and Mary Kane Davis, who came to California in 1852. Joseph farmed and worked as a teamster hauling timber to the mines in Plymouth and Amador City. His descendants still live in the Shenandoah Valley. (Courtesy Mary Cowan.)

Pictured here are Hiram Cloise Farnham and his wife, Eunice Haynes Farnham. Hiram came to California from his native New York in 1850 and settled in Fiddletown in 1853. Here, he built a sawmill and went into the lumbering business in partnership with James McLeod. Hiram and Eunice were married in 1854 and had eight children. Their descendants still live in Amador County. (Courtesy Amador County Archives.)

Andrew Jackson Crain traveled to California from Missouri in September 1849 on the same wagon train that brought other Amador pioneer families like the Bells, Boones, and Hinksons to the Golden State. He was one of the first settlers in the Shenandoah Valley. In 1856, he and Susannah Bell were married and parented 10 children. Andrew worked as the ditch superintendent for the Plymouth Consolidated Mine. (Courtesy Amador County Archives.)

Pictured here are members of the pioneer Uhlinger and Leoni families. From left to right are (first row) Herman Leoni with his dog; (second row) Ursula and Adam Uhlinger, unidentified, and the Uhlingers' son Henry; (third row) Georgia, Steve, and John Leoni and the Uhlinger's daughter. Adam and Ursula planted the first grapevines and opened the first winery in Amador County. (Courtesy Meredith Hansen.)

John McKnight Jameson and his wife, Mary Bacon Jameson, were settlers in the Shenandoah Valley in 1852. John, the son of William and Sarah Jameson of the Calf Pasture River Colony of Virginia, was born there in 1802. The family migrated to Missouri in 1815, and later, John and Mary came to California. Mary's ancestors were associated with the 1676 Bacon's Rebellion in Jamestown, Virginia. (Courtesy Amador County Archives.)

Reuben Ball and his wife, Catherine, were also pioneer settlers in the Shenandoah Valley. Here, descendants of the couple gather for a family photograph. From left to right are Charles W. Ball, Jack Crain, Olive Ball Crain with baby Ralph, Laura Ball, Oliver Ball (son of Reuben and Catherine) holding Raymond, Mabel, Mernelva holding Clyde, Edith, Mattie, Fred, and George Watson Ball. (Courtesy Amador County Archives.)

Ten

21st-Century State of Mind

A Thriving Community

It has been nearly two centuries since Jedediah Strong Smith ventured into the land of the Miwok to become the first Euro-American to explore the upper reaches of the Cosumnes River drainage and nearly 175 years since miners trekked from around the world seeking fortune in the goldfields. Where the adventurous first set foot in the area that was destined to include the northwestern region of Amador County, the face of the land has in many ways changed drastically and in others seems to remain the same.

The last decades of the 19th century saw a boom in mining, the birth of towns, and a proliferation of agricultural interests. Roads to transport goods were laid out, and schools, churches, and clubs were organized. The wild days of the early West gave way to a more settled existence. After the turn of the 20th century, the economy fluctuated, stricken by several hard blows. First it was the Great Depression. Then on December 7, 1942, mines across the country were closed down by a presidential order. Just as the economy began to rebound in the 1950s and 1960s, the country was hit with economic recessions. Although some storefronts lie empty in the towns and several small family farms have struggled to stay afloat, what remains is a dedication to maintain a small-town, rural ambiance in the community.

The late 20th century brought with it a boom in the wine industry. Wineries in the Shenandoah Valley and around Fiddletown draw the tourist trade, while tasting rooms and upscale dining in Plymouth have brought a new vitality to the town. Tourists are also drawn to visit historic sites in Plymouth, Drytown, and Fiddletown and partake in recreational activities on the Cosumnes River. Improvements in infrastructure are bringing Plymouth into the 21st century, while new housing developments and a coming Indian gaming casino will attract more residents and visitors to the area. However, despite these changes, the northwestern region of Amador County remains one of the most sought out destinations in California today.

The Plymouth Veteran's Memorial Park on Highway 49 was established in 2005. In 2018, several groups came together to add improvements and expand the park. Most recently, the City of Plymouth, the Amador County Veterans Association, and the Plymouth-Foothills Rotary began renovation of the park, adding new signage, an irrigation system, flagpole lighting, and a landscaped entryway. (Courtesy Michael Spinetta.)

Keeping up with 21st century traffic can be a tough challenge for the small town of Plymouth and those drawn to visit the world-renowned Shenandoah Valley wineries nearby. Plymouth met this need by working with the Amador County Transportation Commission and Caltrans to build a traffic circle at the east end of Main Street. The project began in 2017, and construction was completed in 2018. (Courtesy City of Plymouth.)

From the historic buildings along Main Street to the pioneer cemetery on top of the hill, Plymouth residents and visitors alike are reminded of the past. Residents take pride in their history and are dedicated to maintaining it. In this photograph, a recently installed pathway in the cemetery allows visitors to navigate more easily. The path was completed as a Boy Scout Eagle project. (Courtesy Michael Spinetta.)

Plymouth Park has long been a gathering place to celebrate various special occasions or just to relax on a warm summer evening. In this photograph, locals gather to visit and patronize the weekly farmer's market. Most recently, small kiosks have been constructed for a pop-up market venue. (Courtesy Michael Spinetta.)

In Plymouth, the education of children has always been of utmost importance. In this photograph, beloved Shenandoah Valley resident and winery owner Laura Spinetta helps children harvest flowers from the Plymouth Elementary School garden. Spinetta, who passed away in 2011, was a dedicated volunteer at the school, establishing a reading program, working with children in the garden, and assisting wherever she was needed. (Courtesy Michael Spinetta.)

Another vital learning tool for children is Plymouth's public library, which was in danger of closing in 2017 when the county was asked to vacate the building it was renting. Then in January 2018, the library reopened at this location on Main Street. There is also a little free library kiosk in Plymouth Park placed by Girl Scout Troop 2063 cadettes. (Courtesy Michael Spinetta.)

The Central Sierra Miwok have long struggled to find their place and thrive in modern society. In 2020, the Bureau of Indian Affairs placed land into federal trust for the Ione Band of Miwok, on which they will build a gaming facility. In this photograph, tribal chairperson and tribal government attorney Sara Dutschke Setshwaelo holds a sign representing a bright future for the tribe. (Courtesy Sara Dutschke Setshwaelo.)

In 2021, when the Caldor Fire burned 221,835 acres in Amador, El Dorado, and Alpine Counties, fire crews and residents answered the need of those impacted. Residents set up a large animal evacuation center at the fairgrounds and delivered food and clothing to evacuated families at the 49er RV Park. As a thank-you to the firefighters, residents painted this mural on a Plymouth fence. (Courtesy Michael Spinetta.)

DISCOVER THOUSANDS OF LOCAL HISTORY BOOKS FEATURING MILLIONS OF VINTAGE IMAGES

Arcadia Publishing, the leading local history publisher in the United States, is committed to making history accessible and meaningful through publishing books that celebrate and preserve the heritage of America's people and places.

Find more books like this at
www.arcadiapublishing.com

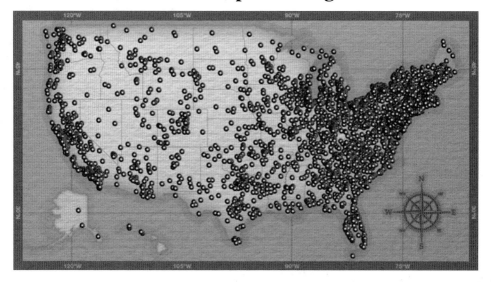

Search for your hometown history, your old stomping grounds, and even your favorite sports team.